D1506919

The
Environment
Encyclopedia

Edited by

Ruth A. Eblen
and
William R. Eblen

11

Reference Volume

Marshall Cavendish
New York • London • Toronto • Sydney

Marshall Cavendish Corporation
99 White Plains Road
Tarrytown, New York 10591-9001

Website: www.marshallcavendish.com

© 2001 Marshall Cavendish Corporation

All rights reserved. No part of this book may be reproduced or utilized in any form or by any means electronic or mechanical, including photocopying, recording, or by an information storage and retrieval system, without prior written permission from the publisher and copyright holder.

Library of Congress Cataloging-in-Publication Data

The environment encyclopedia: acid rain-zoning / edited by Ruth A. Eblen, William R. Eblen.

   p. cm.

   Includes bibliographical references and index.

   ISBN 0-7614-7182-0 (set) — ISBN 0-7614-7183-9 (v. 1) — ISBN 0-7614-7184-7 (v. 2) — ISBN 0-7614-7185-5 (v. 3) — ISBN 0-7614-7186-3 (v. 4) — ISBN 0-7614-7187-1 (v. 5) — ISBN 0-7614-7188-X (v. 6) — ISBN 0-7614-7189-8 (v. 7) — ISBN 0-7614-7190-1 (v. 8) — ISBN 0-7614-7191-X (v. 9) — ISBN 0-7614-7192-8 (v. 10) — ISBN 0-7614-7193-6 (v. 11)

   1. Environmental sciences—Encyclopedias. 2. Ecology—Encyclopedias. 3. Earth sciences—Encyclopedias. I. Eblen, Ruth A. II. Eblen, William R.

GE10.E55   2000
363.7'003—dc21   99-086986

Printed and bound in Italy
06 05 04 03 02 01 6 5 4 3 2 1

Editorial and Production Staff

Project Editors
Debra M. Jacobs Freeman
Joyce L. Tavolacci

Editorial Director
Paul Bernabeo

Manuscript Editors
Michael Casey
Caroline Hutchings Edlund
June English
Susan Gamer
Nancy Gratton

Proofreaders
Dorothy Bauhoff
Nicole Gilliam

Editorial Assistant
Sherry Streicker

Contributors of Special Features
Susan Albury
Karen Ang
Paul Bernabeo
Nicholas Freudenberg
Susan Gamer
Nicole Gilliam
Senta Korb
George Kraemer
Jonathan Latimer
Christopher Manes
Suzanne Rothberg
Douglas Sanders
Joyce L. Tavolacci
William T. Waller

Designers
Howard Petlack, A Good Thing, Inc.
Dianne Stetch-Rubin, A Good Thing, Inc.

Illustrator
Seth Jabour

Photo Researchers
Caroline Casey
Jane Sanders
Joyce L. Tavolacci

Indexer
Cynthia Crippen, AEIOU, Inc.

Color Separation
Embassy Graphics

Production and Manufacturing Director
Michael Esposito

Table of Contents

About This Encyclopedia

This encyclopedia could not have been created without the guidance, dedication, and expertise provided by Ruth A. Eblen, president of the René Dubos Center for Human Environments, and William R. Eblen, the center's founding president. During their more than twenty-five years of leadership in the movement for public literacy about the environment, the Eblens have engineered a framework for gathering, refining, and disseminating positive values about the care and maintenance of the earth from the fields of science, industry, law, government, philosophy, religion, art, and the public interest.

Articles. Four hundred articles were selected and arranged in alphabetical order to represent a broad spectrum of topics appropriate for students and general readers. Earlier versions of many articles appeared previously without illustration in *The Encyclopedia of the Environment*, published by Houghton Mifflin Company in 1994. These were updated, illustrated, and reedited for readers who do not have a strong background in science. Several articles were newly commissioned based on the advice of librarians and members of the teaching community. Articles have been written for the most part by leading specialists on each subject. Effort has been made to simplify language and ideas. However, some articles cover relatively sophisticated subjects deemed by the editors to be necessary for inclusion in any comprehensive reference work designed to promote environmental literacy.

Cross-references. Because every topic in environmental studies is related organically to numerous others, virtually every article ends with a list of closely related articles that can be found elsewhere in this encyclopedia. In addition, several articles include special Linkup boxes that elaborate on topical interrelationships.

Additional Sources. Articles include a list of additional sources in print and electronic media. Volume 11 includes an extensive review of the ever-growing number of resources in the field of environmental studies. This section was written by a librarian with young researchers in mind and is designed to guide and encourage online research.

Photographs. Photographs appear on nearly every page of this encyclopedia. Many of the images selected for inclusion were provided through the auspices of the René Dubos Center. Several of these images were entries in the United Nations Environment Programme (UNEP) International Photographic Competition on the Environment, sponsored by Canon, Inc. Noel Brown, former North American director of UNEP and chairman of the board of trustees of the René Dubos Center, was instrumental in harnessing UN resources for use in this encyclopedia. Researched by Caroline Casey and reviewed by the editors, many of these images depict positive, international efforts to improve the condition of the earth. Other images, selected by in-house editors, illustrate environmental stress and the often heated public debate about the environment.

In addition to the United Nations, especially the UNEP archive at Topham Picturepoint, Kent, England, the editors gratefully acknowledge the resources provided by various other organizations, including the American Cancer Society, Earth First!, and the Sierra Club as well as the following departments and agencies of the U.S. government: Army Corps of Engineers, Digital Visual Library; Centers for Disease Control and Prevention; Dept. of Agriculture; Fish and Wildlife Service; Library of Congress; National Aeronautics and Space Administration.

Graphic Illustrations. Hundreds of charts, graphs, tables, and maps were created especially for this encyclopedia as a means to instruct students about the ways scientists represent numerical data in graphical form. The data have been made available by contributors or drawn from the following sources: British Geological Survey; Energy and Environmental Research Center; Energy Information Agency; Environmental Business International, Inc.; European Environmental Agency; International Geosphere-Biosphere Program (IGBP); National Center for Health Statistics; The René Dubos Center for Human Environments; San Francisco Convention and Visitors Bureau; United Nations; U.S. Army, Soldier Systems Center; U.S. Bureau of the Census; U.S. Centers for Disease Control and Prevention; U.S. Conference of Mayors, Garbage Solutions; U.S. Dept. of Agriculture (USDA); U.S. Dept. of Commerce; U.S. Dept. of Energy; U.S. Dept. of Labor, Bureau of Labor Statistics; U.S. Environmental Protection Agency (EPA); U.S. Geological Survey (USGS); World Conservation Union (IUCN), World Conservation Monitoring Center; World Health Organization (WHO); World Resources Institute; International Institute for Environmental Development; and The Worldwatch Institute.

In addition, the following print sources were consulted: *The Encyclopedia of the Environment* (Houghton Mifflin Company, 1994); *State of the World 1998* (W.W. Norton and Company, 1998); *State of the World 2000* (W.W. Norton and Company, 2000); *Statistical Record of the Environment,* 3rd edition (Gale Research Inc., 1995); *Vital Signs 1999* (W.W. Norton and Company, 1999); and *The World Almanac and Book of Facts 1999* (World Almanac Books, 1998).

Other data were selected from issues of the following journals: *British Petroleum Statistical Review of World Energy* (June 1999); *Environmental Business Journal; Hydrocarbon Processing; International Energy Annual* (April 1999); *Materials Engineering; Oil and Gas Journal;* and *World Oil.*

Reference Volume. Volume 11 includes a variety of additional guides for students, including a time line that charts the expansion of awareness of environmental issues in North America and the world as well as a glossary of more than four hundred technical terms. An extended essay on resources for further research includes valuable guidance for students, teachers, and parents on use of the Internet and other print and audiovisual media.

Seven thematic indexes reflect the conceptual organization of this set of articles. Finally, a comprehensive index completes and augments the brief indexes appearing at the back of each of the first ten volumes.

Environmental Time Line

Prepared by Jonathan Latimer

This time line arranges major events in the history of environmental experience and thinking in chronological order. The events listed here focus on the steady development of awareness and understanding of the environment of North America and, eventually, the world. The time line also reveals a steady increase of concern for what may be lost if the land, air, and water are not treated wisely.

Before 1700

1492—Christopher Columbus lands in the Bahamas Islands and describes the land as "so green that it is a pleasure to gaze upon it."

1632—Thomas Morton publishes *New England Canaan*, which includes a survey of natural resources.

1634—*New England's Prospect* by William Wood describes the land and the Native Americans he encountered.

1672—John Joselyn publishes *New England's Rarities Discovered*, which includes a list of weeds introduced by European colonists into America. Two years later he publishes *An Account of Two Voyages to New England, Made during the Years 1638, 1663*, describing the changes in the land and natural resources, including the decline in the number of wild turkeys.

1678—John Bannister arrives in Virginia, where he collects plants and insects to send back to England. He is sometimes called America's first resident naturalist.

1681—William Penn requires Pennsylvania settlers to preserve one acre of trees for every five acres cleared.

1700s

1709—John Lawson publishes *A New Voyage to Carolina*, the first major natural history of America.

1731–1743—Mark Catsby publishes the first full natural history of America in two volumes entitled *The Natural History of Carolina, Florida, and Bahamas Islands*.

1735—Swedish naturalist Carolus Linnaeus publishes *Systema Naturae*. It lays out the scientific scheme still used today for naming plants and animals.

1748–1762—Jared Eliot, clergyman and physician, writes *Essays on Field Husbandry in New England*, promoting soil conservation.

1762–1769—Philadelphia committee led by Benjamin Franklin attempts to regulate waste disposal and water pollution.

1775–1783—American Revolution

1776—*The Declaration of Independence* published on July 4.

1782—Jean De Crèvecoeur publishes *Letters from an American Farmer*, describing rural life and nature.

1784–1785—Thomas Jefferson describes the natural world around his home in *Notes on the State of Virginia*.

1789—Gilbert White, an obscure clergyman in rural England, publishes his quirky and whimsical book, *The Natural History of Selborne*.

1791—The New York state assembly closes the hunting season on the heath hen. The species is extinct by the early 1900s.

—William Bartram's *Travels* describes the beauty of the American land.

1798—*Essay on the Principle of Population* by Thomas Malthus published in England. In it, he argues that the growth of the human population will outpace humanity's ability to feed itself, leading to famine and warfare over scarce resources. Improvements in farming and preservation techniques and in methods of distribution have prevented Malthus's most dire predictions from coming true.

1799—The American bison becomes extinct in eastern North America when the last one is killed in Pennsylvania.

1800s

1803—Louisiana Purchase doubles the size of the United States.

1804—Lewis and Clark Expedition leaves to explore the lands of the Louisiana Purchase and find a route to the Pacific Ocean.

1808—First volume of Alexander Wilson's *American Ornithology* published. When completed in 1814, the study fills nine volumes with descriptions of 260 species.

1812–1814—War of 1812

1818—Massachusetts bans the hunting of robins and horned larks, both popular food items.

1823—James Fenimore Cooper writes *The Pioneers*, which contains the idea that humans should "govern the resources of nature by certain principles in order to conserve them."

1824—The Society for the Prevention of Cruelty to Animals (SPCA) founded. Launched in Britain, this was one of the first organizations devoted to protecting domesticated animals and, later, wild animals.

1825—The Erie Canal opens, providing access between the East and the Great Lakes and Midwest.

1826—James Fenimore Cooper publishes *The Last of the Mohicans*.

1826–1838—John James Audubon publishes *The Birds of North America*.

1831–1839—Audubon publishes his *Ornithological Biography*.

1832—Arkansas Hot Springs established as a national reservation, setting a precedent for protecting Yellowstone in Wyoming and, eventually, establishing the national park system.

—Thomas Nuttall publishes *A Manual of the Ornithology of the United States and Canada* describing the interrelationship between behavior and habitat in birds.

1836—Ralph Waldo Emerson publishes his essay "Nature."

1845—Henry David Thoreau moves to Walden Pond.

—Death of Johnny Appleseed, who planted apple trees across Ohio and Indiana for nearly 50 years.

1847—A Fourth of July speech by George Perkins Marsh calls attention to the destructive impact of deforestation and advocates a conservationist approach to managing forests. The speech becomes the basis for his book *Man and Nature*, published in 1864.

1849—Gold Rush to California

1851—Thoreau delivers his lecture *The Wild* for the first time.

—Herman Melville publishes *Moby-Dick*.

1852—Mother of the Forest, a giant sequoia tree, chopped down in the Calaveras Grove of Big Trees in California. This site eventually becomes Yosemite National Park.

1854—Henry David Thoreau publishes *Walden*, his account of living alone in the New England countryside, in which he expresses his disillusion with urbanization and industrialization.

1857—State of Vermont commissions George Perkins Marsh to study the depleted fish populations in the Connecticut River.

—Frederick Law Olmsted begins developing New York City's Central Park.

1859—Charles Darwin publishes *The Origin of Species.*

1861–1864—The American Civil War

1864— George Perkins Marsh publishes *Man and Nature; or, The Earth as Modified by Human Action.* It is the first comprehensive study of the human impact on the environment and presents Marsh's theory that the decline of cultures is a result of their abuse of the environment. The book stresses forest preservation and soil and water conservation as ways of preventing social decline.
    —Posthumous publication of Henry David Thoreau's *The Maine Woods,* in which he calls for the establishment of "national preserves" of virgin forest.

1866—The word *ecology* coined by the German biologist Ernst Haeckel.
    —The American Society for the Prevention of Cruelty to Animals (ASPCA) founded.

1867—Alaska purchased by the U.S. government.

1869—John Wesley Powell leads expedition down the Colorado River through the Grand Canyon.
    —John Muir moves to Yosemite Valley.
    —The Transcontinental Railway completed with the driving of the Golden Spike at Promontory Point, Utah.

1871—*Wake-Robin,* John Burroughs's first book of essays on nature, published.

1872—Congress passes legislation making Yellowstone the world's first official national park.
    —Geologist Clarence King publishes *Mountaineering in the Sierra Nevada.*

1873—Henry Winchester invents the repeating rifle.

1876—Appalachian Mountain Club founded.
    —American Forestry Association campaigns to cut timber on government reserves. The American Association for the Advancement of Science calls for federal legislation to protect timberlands.

1878—The Labrador Duck becomes extinct when the last one is killed on Long Island.
    —Barbed wire introduced in the American West.

1879—U.S. Geological Survey formed.

1883—The last bison hunt held.

1885—U.S. Biological Survey created because of concern about the decline in the populations of the buffalo and the passenger pigeon.

1886—First Audubon Society formed by George Bird Grinnell.

1890—General Federation of Women's Clubs founded. Conservation is one of the organization's major concerns.
    —Yosemite National Park and Sequoia National Park established in California.

1891—Congress passes the Forestry Reserves Act, giving the president the power to create "forest reserves." This becomes the legislative foundation for the national forest system.

1892—Sierra Club founded with John Muir as president.

1893—President Benjamin Harrison creates 13 million acres of forest reserves, including 4 million acres covering much of the High Sierra.
    —Britain's National Trust set up to purchase land in order to preserve natural beauty and cultural landmarks.
    —Gasoline-powered internal combustion engine invented.

1894—*The Mountains of California,* John Muir's first book, published.

1899—Congress passes the Rivers and Harbors Act, which authorizes the U.S. Army Corps of Engineers to investigate and prohibit the dumping of wastes into navigable waterways.
    —*Bird-Lore* magazine, known later as *Audubon,* launched by Frank M. Chapman.

1900s

1900—Wild buffalo population drops to fewer than 40 animals.
    —The automobile is welcomed as a relief from pollution. New York City, with 120,000 horses, must dispose of 2.4 million pounds of manure every day.
    —Congress passes the Lacey Act, prohibiting the shipment of birds killed in violation of laws of any state or foreign country.
    —The last wild passenger pigeon killed in Ohio.

1901—President William McKinley assassinated. Theodore Roosevelt becomes president of the United States.

—First Sierra Club outing held at Tuolumne Meadows in Yosemite.

1903—Theodore Roosevelt visits Yosemite with John Muir.

—Theodore Roosevelt creates first national wildlife refuge on Pelican Island, Florida.

—First manned flight in a powered airplane by Orville and Wilbur Wright.

— Mary Austin publishes *The Land of Little Rain*, describing the desert landscape of California and Arizona and the people who live there.

—Novelist Jack London publishes *The Call of the Wild*.

1904—The last Carolina parakeet seen in the wild.

1905—National Audubon Society organized.

—U.S. Forest Service created as separate bureau.

—Gifford Pinchot becomes head of the U.S. Forest Service. Pinchot believes in managing natural resources for the benefit of humanity. A battle between Pinchot and John Muir, the leading voice for preserving wilderness, develops over building a dam and flooding the Hetch Hetchy Valley in Yosemite.

1906—Congress passes the Food and Drugs Act.

—The first edition of *The Writings of Henry David Thoreau* published in 20 volumes.

—Jack London publishes *White Fang*.

1909—National Conservation Commission suggests "broad plans . . . be adopted providing for a system of waterway improvement."

—Charles Van Hise writes *The Conservation of Natural Resources*.

—U.S.–Canada Boundary Pollution Commission established.

—The presidency of Theodore Roosevelt ends. During his administration, the government has set aside 42 million acres of national forests, 51 national wildlife refuges, and 18 areas of special interest, including the Grand Canyon.

1910—Gifford Pinchot publishes *The Fight for Conservation*.

1912—The U.S. establishes the first national park system in the world. Today, there are thousands of national parks in more than 120 countries.

—National Audubon Society begins campaign to boycott hat makers using the feathers of endangered tropical birds.

1913—Migratory Bird Act prohibits hunting and marketing migratory birds in spring. Act also prohibits importing wild bird feathers for women's fashion.

—William T. Hornaday, head of New York Zoological Society, writes *Our Vanishing Wildlife, Its Extermination and Preservation*. He forms the Wildlife Protection Fund with financial support from Andrew Carnegie, Henry Ford, and George Eastman.

—Hetch Hetchy Dam in Yosemite National Park approved by Congress over strong objections by John Muir and the Sierra Club.

1914—Martha, the last known passenger pigeon, dies at the Cincinnati Zoo.

1914–1917—World War I

1915—Congress creates Rocky Mountain National Park.

—California legislature authorizes $10,000 to start planning and constructing the John Muir Trail in the Sierra Nevada Mountains.

1916—Congress creates the Office of National Parks, Buildings, and Reservations (renamed the National Park Service in 1934).

1918—Congress ratifies the Migratory Bird Treaty.

—Incas, the last known Carolina parakeet, dies at the Cincinnati Zoo.

1919—Acadia National Park created in Maine.

—National Parks and Conservation Association founded.

1920—Mineral Leasing Act opens up rich deposits on federal lands for token rental fees.

—Water Power Act authorizes federal hydroelectric projects.

—John Burroughs publishes *Accepting the Universe*, arguing that personal experience is equal to science.

1921—General Motors researchers discover tetraethyl lead, a gasoline additive that improves the performance of automobile engines. The new fuel, called ethyl leaded gasoline, goes on sale without safety tests.

1923—Izaak Walton League founded. It lobbies in Washington, D.C., to fight a Mississippi Valley dredging project.

1924—Five workers at a Standard Oil refinery making ethyl lead gasoline die "violently insane."
—Aldo Leopold helps establish the first wilderness reserve within a National Forest in New Mexico.
—First Audubon Sanctuaries open in Louisiana and Long Island.

1925—Five more workers die in a New Jersey DuPont plant that makes tetraethyl lead.

1926—First large-scale survey of air pollution in United States takes place in Salt Lake City.
—A committee of experts directed by the surgeon general permits ethyl leaded gasoline to be sold. Public health experts protest, but leaded gasoline stays on the market until 1986.

1927—Five New Jersey women, dying from radium poisoning, file lawsuits against U.S. Radium Corporation. Radium widely used to make luminous wristwatch faces and other instruments.

1928—Radium lawsuits settled out of court.
—Public Health Service begins checking air pollution in the eastern United States. They find that smoke and soot cut sunlight by 20 to 50 percent in New York City.
—Henry Beston publishes *The Outermost House*, describing a year spent on the dunes at the tip of Cape Cod, Massachusetts.

1929—Over 100 wildlife sanctuaries are brought under federal protection by the Norbeck-Anderson Act.
—Start of the Great Depression.

1930—Discovery of chlorofluorocarbon gases (CFCs). They are widely used in refrigeration, packaging materials, and in aerosol sprays that propel everything from paint to deodorant.

—Food and Drug Administration establishes new standards of quality and adopts its present name.
—Carlsbad Caverns becomes a National Park.

1932—The last heath hen seen in the wild.

1933—Civilian Conservation Corps (CCC) formed. Volunteers plant trees, build roads, and construct thousands of fire towers, buildings, and bridges on federal land.
—Tennessee Valley Authority (TVA) established to construct dams and waterworks to control flooding and create electric power for the region.

1934—Dust bowl storms erode topsoil on the Great Plains. One storm alone blows away more than 350 million tons of topsoil.
—Lewis Mumford publishes *Technics and Civilization*, examining the environment of cities.
—Roger Tory Peterson publishes *A Field Guide to the Birds* and revolutionizes bird watching.

1935—Soil Conservation Service established.
—Wilderness Society founded by Aldo Leopold and Arthur Carhardt.
—William Vogt becomes editor of *Bird-Lore* magazine with Roger Tory Peterson as art director.
—*An Almanac for Moderns* by Donald Culross Peattie appears.

1935–1955—Bernard Devoto writes a controversial monthly column called "The Easy Chair" for *Harper's* magazine. Many columns are concerned with conservation, especially the protection of public lands in the West.

1936—National Wildlife Federation formed.

1937—A survey of air pollution in New York City shows that conditions are worse than they were in 1928.

1937—Pittman-Robertson Act sets up an excise tax on sporting arms and ammunition. The proceeds are to be used for projects to protect wildlife.
—Ducks Unlimited formed by hunters to protect migrating waterfowl.

1938—Ernest Hemingway publishes *The Snows of Kilamanjaro*.

1939—U.S. Fish and Wildlife Service created.

    —An episode of choking smog in Saint Louis induces people throughout the United States to switch from soft coal to hard coal or fuel oil.

1940—Bald Eagle Protection Act signed.

1941—Saint Louis adopts first strict smoke control ordinance in the United States.

    —John Steinbeck publishes *The Sea of Cortez*. Written with marine biologist Edward F. Rickets, it reports on a research voyage to the Gulf of California.

1941–1945—World War II

1942—First atomic chain reaction takes place at the University of Chicago.

    —William Faulkner publishes *The Bear*.

1944—Big Bend National Park established in Texas.

    —*The Wolves of Mount McKinley* published. It is the first scientific study of wolves in the wild.

    —Sally Carrighar publishes *One Day at Beetle Rock*, a detailed narrative of events that occur during 24 hours at a granite outcrop in the Sierra Nevada Mountains.

1945—U.S. Corps of Engineers abandons the Potomac River dam project after protests from the Izaak Walton League and the National Parks Association.

    —President Harry Truman issues the Proclamation on the Continental Shelf, which clears the way for offshore drilling for oil.

    —A project to protect the endangered whooping crane started.

1946—International Whaling Commission founded.

1947—Los Angeles Air Pollution Control District formed. It is the first air pollution control department in the nation.

    —Defenders of Wildlife founded.

    —Everglades National Park created in Florida.

1948—Congress passes Federal Water Pollution Control Act.

    —A smog disaster in Donora, Pennsylvania, near Pittsburgh, leaves 20 people dead and 600 hospitalized.

    —A killer fog in London is blamed for 600 deaths.

    —Fairfield Osborn publishes *Our Plundered Planet*, challenging the belief that human technology can create universal well-being.

    —In *Road to Survival*, William Vogt warns that the uncontrolled growth of the human population will lead to starvation and social upheaval.

    —The International Union for the Conservation of Nature (IUCN) founded.

    —Paul H. Müller wins the Nobel Prize for Medicine for inventing the pesticide DDT.

1949—*Sand County Almanac* by Aldo Leopold published posthumously. It contains Leopold's classic formulation of a "land ethic," which recognizes that other species have intrinsic value and are not just meant to be exploited by humans.

    —First national conference on air pollution sponsored by the Public Health Service.

    —Izaak Walton League issues "Crisis Spots in Conservation," identifying specific water projects it opposes.

1951—The Nature Conservancy formed.

    —Rachel Carson publishes *The Sea around Us*. It becomes a best-seller and introduces the natural history of the ocean to a wide audience.

1952—Three to four thousand people die during killer fogs in London. The smog is so thick that buses must be escorted by men walking with lanterns on the streets.

    —David Brower becomes the first executive director of the Sierra Club.

    —Joseph Wood Krutch publishes *The Desert Year*.

1953—A smog incident in November kills between 150 and 250 in New York City.

1954—Heavy smog shuts down industry and schools in Los Angeles for most of October.

    —Justice William O. Douglas of the U.S. Supreme Court leads a 189-mile hike to protest a proposed highway that would replace the scenic Chesapeake and Ohio Canal. The plan is dropped because of the publicity.

—Joseph Wood Krutch publishes *The Voice of the Desert*. His examination of the adaptations of life to the desert environment leads Krutch to reconsider humanity's place in the world.

1955—Congress passes Air Pollution Research Act.

—International Air Pollution Congress held in New York City.

—As a result of public pressure, the U.S. government drops plans for the Echo Park Dam in Dinosaur National Monument.

1956—Congress passes Water Pollution Control Amendments.

1957—Loren Eiseley publishes *The Immense Journey*.

1958—First Public Health Service National Conference on Air Pollution.

—Sigurd Olson publishes *The Singing Wilderness*.

1959—California becomes the first state to impose standards for automotive emissions.

—The opening of the St. Lawrence Seaway provides access for ships from the Great Lakes to the Atlantic Ocean.

—Peter Matthiessen publishes *Wildlife in America*. It is a survey of animal extinctions and efforts at protection across the United States.

1960—The Public Health Service begins a two-year study on how automobiles contribute to air pollution.

—Wallace Stegner writes *The Wilderness Letter*, which is instrumental in establishing the National Wilderness Preservation System four years later.

—Justice William O. Douglas leads hikes and writes articles and books in support of conservation, including *My Wilderness: The Pacific West*.

1961—International Clean Air Congress held in London.

—World Wildlife Fund founded.

1962—Rachel Carson publishes *Silent Spring*, alerting the general public to the danger of pesticides, especially DDT, to wildlife and humans. Widely read and debated, the book is attacked by scientific, business, and political authorities. Over time, her main argument is overwhelmingly confirmed.

—Barbara Ward examines threats to the global environment in *The Rich Nations and the Poor Nations*.

—Hal Borland publishes *Beyond Your Doorstep: A Handbook of Country Living* about the joys of country life and the natural world.

1963—Congress passes the Clean Air Act with $95 million for study and cleanup efforts at local, state, and federal levels.

—A massive fish and wildlife kill takes place in the lower Mississippi River. Its cause is traced to DDT.

—John Hay publishes *The Great Beach*.

—Farley Mowat publishes *Never Cry Wolf*, one of the first books to show wolves in a favorable light.

1964—Congress passes the Wilderness Act, creating a national wilderness preservation system and establishing a process for the permanent protection of certain land from development.

1965—Congress passes the Water Quality Act, setting standards for clean water.

—The Sierra Club brings suit to protect New York's Storm King Mountain from a power project. The court allows the club to participate in the case, even though it does not represent an economic interest, establishing an important precedent.

—Weather inversion creates four days of unusually strong air pollution in New York City; 80 die.

—Sierra Club, led by David Brower, conducts an effort to save Grand Canyon from dams proposed by the Bureau of Land Management.

—First photo of the earth from space taken from the Lunar Orbiter.

—Edwin Way Teale publishes *Wandering through Winter*, which wins the Pulitzer Prize in 1966.

1966—Ecologist Barry Commoner calls attention to the threats to the environment posed by technological excesses in *Science and Survival*.

1967—Environmental Defense Fund formed on Long Island to fight the use of DDT.

—The supertanker *Torrey Canyon* spills thousands of gallons of oil off Cornwall, England. Large quantities float ashore, polluting beaches and killing wildlife.

—Fund for Animals founded.

—Lynn White Jr. writes the essay "The Historical Roots of our Ecological Crisis." He argues that the idea of human mastery over nature has deep historical sources in the Judeo-Christian tradition.

1968— A plan to build a dam in Grand Canyon stopped.

—Congress protects wild areas by passing the Wild and Scenic Rivers Act and the National Trails Act.

—Paul Ehrlich publishes *The Population Bomb*. Building on the arguments of Thomas Malthus and William Vogt, Ehrlich's polemic stirs the first major debate in modern times about the consequences of human population growth.

—Zero Population Growth (ZPG) founded. The group promotes limiting human population growth to a level where births and deaths occur at the same rate.

—Garrett Hardin publishes *The Tragedy of the Commons*. This essay points out that the cumulative effect of little everyday decisions can be ruinous in the long term. It argues that developing a longer range view is crucial to the equitable sharing of resources in order to sustain humanity.

—Edward Abbey publishes *Desert Solitaire*, a sharply satiric but very humorous examination of humanity's relationship with wilderness and wildness.

1969—David Brower, who has been ousted from the Sierra Club, founds Friends of Earth (FoE).

—Santa Barbara oil spill from offshore wells pollutes beaches in southern California, arousing public anger against polluters.

—Union of Concerned Scientists founded.

—Cuyahoga River near Cleveland, Ohio, bursts into flame.

—René Dubos publishes *So Human an Animal;* Wendell Berry, *The Long-Legged House;* John Hay, *In Defense of Nature;* Edward Hoagland, *Notes from a Century Before;* Gary Snyder, *Earth Household;* John and Mildred Teale, *Life and Death of a Salt Marsh;* and Paul Shepard, *Subversive Science.*

1970—Congress passes Clean Air Act Amendments, which greatly expand protections against air pollution.

—National Environmental Policy Act (NEPA) passed.

—The Environmental Protection Agency created.

—League of Conservation Voters formed.

—First Earth Day held on April 22. Twenty million people take part in events in over 2,000 communities across the United States.

—Natural Resources Defense Council created.

—Aldo Leopold's *Sand County Almanac* (1949 reissued.

1971—Greenpeace founded.

—Edward Hoagland publishes *The Courage of Turtles.*

—Dr. Seuss gives a warning about greed and environmental destruction in his cautionary tale *The Lorax.*

1972—Congress passes Federal Water Pollution Control Act, Coastal Zone Management Act, and the Ocean Dumping Act.

—Environmental Protection Agency bans DDT, a victory for the Environmental Defense Fund and the National Audubon Society.

—United Nations Conference on the Human Environment held in Stockholm, Sweden. Called the "Only One Earth" conference, it pulls together and publicizes many of the strands of environmental concern. The conference leads to the creation of the United Nations Environment Programme (UNEP).

—René Dubos and Barbara Ward publish *Only One Earth: The Care and Maintenance of a Small Planet*, an outline of plans for managing the earth.

—First bottle-recycling bill passes in Oregon.

—The Sierra Club wins its case against the Walt Disney Company when the Supreme Court limits the commercial development of Mineral King Valley in California.

—A team from the Massachusetts Institute of Technology uses advanced computer techniques to estimate what could happen if the human population continues to grow and consume resources at present rates. Its report, *Limits of Growth*, published by the Club of Rome, sparks much debate. The predictions are alarming, but the analysis necessarily simplifies reality and is unable to quantify factors like human ingenuity. The study is updated in 1992 and published as *Beyond the Limits* by D. Meadows et al.

1973—Congress passes the Endangered Species Act.

—Publication of *Small Is Beautiful* by E. F. Schumacher popularizes the idea that big is not always better.

—Eighty nations sign the Convention on International Trade in Endangered Species (CITES), limiting the commercial exploitation of rare wild animals.

—Arab oil embargo panics U.S. and European consumers; prices quadruple despite the fact that no real shortage exists.

—Tellico Dam controversy breaks out when the Endangered Species Act is used to prevent the building of a dam in order to protect a species of small fish called the snail darter.

—Cousteau Society founded.

1974—Annie Dillard publishes *Pilgrim at Tinker Creek*; Gary Snyder, *Turtle Island*; and Lewis Thomas, *The Lives of a Cell*.

—The United Nations Population Conference in Bucharest stresses the connection between population factors and economic development.

—First warning that human-made chlorofluorocarbons (CFCs) are depleting the protective ozone layer in the upper atmosphere.

—Congress passes Safe Drinking Water Act.

1975—Atlantic salmon return to the Connecticut River after an absence of over 100 years.

—The Worldwatch Institute founded to analyze global environmental problems.

—Edward Abbey publishes the novel *The Monkey Wrench Gang* in which he coins the term *monkeywrenching* to describe acts of sabotage to protect the environment.

—Ernest Callenbach publishes *Ecotopia*.

1976—National Academy of Science warns that CFCs are damaging the ozone layer.

—Congress passes Resource Conservation and Recovery Act to regulate hazardous waste and garbage.

1977—Congress passes Clean Air Act Amendments, which create new deadlines for compliance with emissions standards.

—Congress passes Clean Water Act, setting national standards for pretreatment of industrial waste.

—U.S. Department of Energy created.

—North Sea oil spill releases over 8 million gallons of oil into the ocean.

—New York City blacked out by an electrical power failure.

—Wendell Berry publishes *The Unsettling of America*, depicting life on a small farm and defending rural values.

—Love Canal, a neighborhood in Niagara Falls, New York, found to be built on a former waste dump now leaking toxic chemicals.

1978—After discovery of hazardous waste in the soil around homes and the neighborhood school, President Jimmy Carter declares an emergency at Love Canal and the town is evacuated. Events at Love Canal alert the country to the hidden dangers of pollution in soil and groundwater.

—Barry Lopez publishes *Of Wolves and Men*, which investigates wolves in the wild and examines how humans have depicted wolves in myths and stories.

—Peter Matthiessen publishes *The Snow Leopard*, an extended meditation on wildness during a trek in the mountains of Nepal.

1979—An accident at a nuclear power plant at Three Mile Island, Pennsylvania, almost leads to a meltdown.

—James Lovelock publishes *Gaia: A New Look at Life on Earth*. Lovelock theorizes that the earth as a whole is a self-regulating entity unconsciously maintaining optimal conditions for life.

1980—Congress passes Comprehensive Environmental Response, Compensation, and Liability Act ("Superfund") to create national inventory of hazardous waste sites and to regulate cleanup.

—The passage of the Alaska National Interest Lands Conservation Act adds more than 100 million acres to the federal wild land system.

—A last-ditch effort to save the California condor from extinction begins. All condors remaining in the wild are captured and held in zoos for a captive breeding program. Young condors raised at the zoos will be released back into the wild.

—Earth First! formed.

—*The Global 2000 Report to the President*, prepared for Jimmy Carter, examines the political aspects of dealing with environmental problems.

—In *The Death of Nature: Women, Ecology, and the Scientific Revolution*, Carolyn Merchant explores connections between women's issues and ecology.

1981— John McPhee publishes *Basin and Range*, the first of a series of books about geologists and the geology of North America. It is followed by *In Suspect Terrain* (1983), *Rising from the Plains* (1986), and *Assembling California* (1993), and finally collected as *Annals of the Former World* (1998).

1982—*Nature and Madness*, published by Paul Shepard, traces the roots of madness to humanity's estrangement from nature.

—Earth Island Institute founded.

—Gary Paul Nabhan publishes *The Desert Smells Like Rain*.

1983—Times Beach, Missouri, evacuated when poisonous chemicals, known as dioxins, are discovered in the soil. The federal government offers to buy the whole town.

1984—A toxic chemical cloud accidentally released from a factory in Bhopal, India, kills 2,000 and injures thousands more.

—First *State of the World* report published by Worldwatch Institute. This report on global environmental issues is published annually thereafter.

1985—A team of British scientists finds a hole in the atmospheric ozone layer above Antarctica. The hole allows dangerous amounts of ultraviolet radiation to reach the surface of the earth.

—Greenpeace ship *Rainbow Warrior* sabotaged by French agents at Auckland, New Zealand.

—Gretel Ehrlich gives compelling descriptions of life on Wyoming ranches in *The Solace of Open Spaces*.

1986—A radioactive cloud released by an accident at a nuclear reactor at Chernobyl, Ukraine. More than 450,000 people evacuated from their homes.

—Barry Lopez publishes *Arctic Dreams*, which examines how our minds shape what we perceive.

1987—Thirty-six countries sign the Montreal Protocol agreeing to protect the ozone layer by limiting the production of CFCs.

—The Brundtland Report, *Our Common Future*, released by the United Nations.

—The last wild California condor taken from the wild for a captive breeding program. The program is successful, and by the late 1990s condors are being returned to the wild.

1988—First pictures of forest fires on earth are taken from space by the space shuttle *Discovery*.

—Concern about global warming as a result of the greenhouse effect grows as new evidence accumulates.

—A coalition of developers launches the Wise Use movement to "eradicate the environmental movement" and repeal all limits on exploiting protected lands.

—The first condor chick hatched in captivity.

1989—The supertanker Exxon *Valdez* spills thousands of gallons of oil into Prince William Sound in Alaska. Despite efforts to clean it up, thousands of birds and other animals are killed.

—Bill McKibben examines the dangers of global warming in *The End of Nature*.

1990—Clean Air Act Amendments establish new, stricter auto emissions standards.

—Some 200 million people in 141 countries celebrate the twentieth Earth Day.

—A large oil spill takes place at Huntington Beach, California.

—Norman Maclean publishes the novella *A River Runs through It*.

1991—Iraqi soldiers, retreating from the allies in the Persian Gulf War, set huge oil fires in Kuwait. The fires are so large they are seen and photographed by the crew of the space shuttle in outer space.

—Thirty-nine countries sign the Madrid Protocol, prohibiting mining in Antarctica.

—Rick Bass publishes *Winter*.

1992—The Earth Summit meets in Rio de Janeiro, Brazil. Officially called the United Nations Conference on Environment and Development, it focuses on global environmental changes and sustainable development.

—The gray wolf removed from the list of endangered species.

—The United States bans tuna caught in nets that kill dolphins.

—Terry Tempest Williams publishes *Refuge: An Unnatural History of Family and Place*, a record of the simultaneous tragedies of her mother's death from cancer and the destruction of wildlife at a bird sanctuary.

—First issue of the annual *Vital Signs* published by Worldwatch Institute.

—Edward O. Wilson warns of impending ecological disaster caused by the widespread destruction of ecological systems in *The Diversity of Life*.

—Al Gore publishes *Earth in the Balance* while running for vice president of the United States.

1994—The United Nations conference on Population and Development meets in Cairo to discuss family planning.

—The International Whaling Commission bans whaling in the waters around Antarctica.

—The status of the bald eagle in the lower 48 states changed from endangered to threatened.

—The Georges Banks in the Atlantic Ocean are closed to commercial fishing by the United States and Canada to allow fish populations to recover from overfishing.

1995—The gray wolf reintroduced to Yellowstone Park.

—Ann Zwinger publishes *Downcanyon*.

—Kirkpatrick Sale calls for abandoning technological advances in favor of a more human-centered world in *Rebels against the Future*.

1996—The Convention on Trade of Endangered Species (CITES) starts monitoring the trade in sharks throughout the world.

—David Quammen reports on efforts to understand extinction and preserve biodiversity in *The Song of the Dodo*.

—In his book *The Abstract Wild*, Jack Turner attacks zoos and all other people who work for preservation. He proposes a new conservation ethic that focuses less on preserving things and more on "leaving things be."

1997—Kyoto Protocol commits industrial countries to reduce their emissions of greenhouse gases such as carbon dioxide, methane, and CFCs. Industrial countries also agree to help developing countries avoid creating new emissions as their economies grow.

1998—Congress passes legislation to clean up the Salton Sea in California.

—Images from the Hubble Space Telescope profide evidence for existence of a planet outside the solar system.

—John McPhee publishes *Annals of the Former World*, a collection of his writings about the geology of North America.

1999—Severe drought and unusually high temperatures occur in the West.

—Evan Eisenberg publishes *The Ecology of Eden*, an examination of humanity's place in nature.

—The bald eagle is removed from the U.S. endangered species list.

2000—Measurements of the concentration of CFCs in the atmosphere are lower, due in large part to the limitations on the use of CFCs set by the Montreal Protocol of 1987.

—Despite the reduction in emissions of CFCs, the hole in the ozone layer over the Antarctic is the largest ever measured. Scientists suspect that the effects of global climate change may be making the problem worse.

—Scientists visiting the North Pole find ice-free open water. This has not been seen before in human memory. At the same time, a ship travels through the Arctic Ocean from the west coast of Alaska to eastern Newfoundland without encountering pack ice. Winds and tides move ice around, but scientists suspect that this melting may be due to global warming.

—Representatives from more than 150 nations meet at the Hague in the Netherlands to work on the details for implementing the 1997 Kyoto Protocol on greenhouse gas emissions.

Glossary

abatement To lower, depreciate, or reduce. Often refers to emission reductions in air quality control.

ablation The weathering of a glacier by surface melting; or weathering of rock through the eroding action of water.

absorption The uptake of water or dissolved chemicals by a cell or an organism, e.g., tree roots absorb dissolved nutrients in the soil.

acceptable daily intake (ADI) Largest amount of chemical to which a person can be exposed on a daily basis without resulting in adverse effects.

accretion The addition of air particles to hydrated drops, such as snow, rain, or sleet, by coagulation as the drops fall through the sky.

acid rain Acid deposition is a complex chemical and atmospheric phenomenon that occurs when emissions of sulfur and nitrogen compounds and other substances are transformed by chemical processes in the atmosphere, then deposited on earth. The wet forms, called acid rain, fall as rain, snow, or fog. The dry forms are acidic gases or particles. See encyclopedia entry.

acre-foot A volume of water that covers one acre to a depth of one foot, or 43,560 cubic feet.

adaptation A change that enhances an organism's ability to cope with its environment. See encyclopedia entry.

aeration The act of mixing a liquid with air.

aerobic A biological process that occurs in the presence of oxygen.

air quality standards The level of selected pollutants set by law that may not be exceeded in outside air. Used to determine the amount of pollutants that may be emitted by industry.

albedo Reflection power or the portion of solar radiation that reaches the surface of the earth and is immediately reflected back into the atmosphere.

algae bloom Increase in population of phytoplankton in response to changes in environmental conditions.

allele One of several forms of the same gene.

ambient Any unconfined portion of the atmosphere; open air; outside surrounding air.

American National Standards Institute (ANSI) National organization that coordinates development and maintenance of consensus standards and sets rules for fairness in their development.

anadromous Refers to fish that hatch in freshwater, migrate to the ocean, and return to freshwater only to spawn (lay their eggs).

anaerobic Refers to biological processes that occur in the absence of oxygen.

anoxic Refers to a lack of oxygen; the inadequate oxygenation of the blood; water that has lost oxygen due to the bacterial decay of organic matter.

anthropogenic Refers to something originating from humans and to the impact of human activities on nature.

aquifer A water-bearing layer of rock (including gravel and sand) that will yield water in usable quantity to a well or spring. See encyclopedia entry.

artesian well A well or spring that taps groundwater that is under pressure so that water rises without pumping.

arthropod The animal phylum including crustaceans, spiders, mites, centipedes, insects, and related forms. See encyclopedia entry "Insects and Related Arthropods".

asbestos A mineral fiber that can pollute air or water and cause cancer or asbestosis when inhaled. See encyclopedia entry.

ash Nonorganic, nonflammable substance left over after combustible material has been completely burned. See encyclopedia entry.

assimilative capacity The amount of pollution a body of water can receive without degradation.

atmosphere The sum total of all the gases surrounding the earth, extending about 250 miles (480 km) above the surface. See encyclopedia entry.

atmospheric deposition The contribution of atmospheric pollutants or chemical constituents to land or water ecosystems.

autotrophic A species or ecosystem sustained entirely by food created within the system, e.g., a green plant or an estuary that is a net exporter of organic matter.

bactericide A pesticide used to control or destroy bacteria.

barrage A dam built across the lower reaches of a river used to divert water for irrigation.

barrel A unit of measure for crude oil and petroleum products equivalent to 42 U.S. gallons.

benthic organism Any of a diverse group of aquatic plants and animals that live on the bottom of marine and fresh bodies of water.

Berlin Mandate Establishes a process that enables countries to take appropriate actions against greenhouse gas emissions beyond the year 2000.

bioaccumulation (biomagnification) A process in which chemicals are retained in fatty body tissue and increase in concentration over time. Biomagnification is the increase in tissue accumulation in species higher in the food chain as contaminated food species are eaten.

bioassay A method of testing a material's effects on living organisms. See encyclopedia entry.

biochemicals Chemicals that are either naturally occurring or identical to naturally occurring substances, e.g., hormones, pheromones, and enzymes. Biochemicals function as pesticides through nontoxic, nonlethal modes of regulating growth or by acting as repellents.

biodegradable The ability of a substance to be broken down physically or chemically by microorganisms. For example, many chemicals, food scraps, cotton, wool, and paper are biodegradable; plastics and polyester generally are not.

biodiversity The number and variety of different organisms in the ecological complexes in which they naturally occur. Organisms range from complete ecosystems to the biochemical structures that are the molecular basis of heredity. See encyclopedia entry.

biogenic Refers to creation by biological processes.

biomass The matter in the biosphere that is contained in living organisms.

biome A major portion of the living environment of a particular region characterized by its distinctive vegetation and maintained by local climatic conditions. See encyclopedia entry.

bioremediation The use of living organisms (e.g., bacteria) to clean up oil spills or remove other pollutants from soil, water, and wastewater.

biosphere The volume that includes the lower part of the troposphere (as high as living organisms can fly or be lofted) and the surface of the earth, including the oceans and all living matter. See encyclopedia entry.

biota All of the organisms, including animals, plants, fungi, and microorganisms, found in a given area. See encyclopedia entry.

biotechnology Any technology that is applied to living organisms. See encyclopedia entry.

biotic Refers to any aspect of life, especially to characteristics of entire populations or ecosystems.

btu (British Thermal Unit) Standard unit of measure for the quantity of heat required to raise the temperature of 1 pound of water by 1 degree Fahrenheit.

buttress dam A type of dam made of reinforced masonry or stonework built against concrete.

by-product Materials, other than the intended product, generated as a result of an industrial process.

canopy Layer of vegetation elevated above the ground, made of tree branches and epiphytes.

cap A combination of clay soil and synthetic liner placed over a landfill during closure. The cap serves to minimize leaks during biodegradation of the waste by keeping precipitation out, odors down, and animals away.

carbon 14 An isotope of carbon 12 (containing two more neutrons) that is radioactive and is used in carbon dating.

carbon cycle A complex cycle that circulates carbon through the atmosphere, oceans, and land, as well as plants and animals. See encyclopedia entry.

carbon monoxide (CO) A toxic, odorless, colorless gas produced during fossil fuel or biomass burning. See encyclopedia entry.

carbon sinks Carbon reservoirs and conditions that take in and store more carbon than they release, such as forests and oceans.

carbonyl sulfide (COS) A gas that is very stable and unreactive in the troposphere but which photolyzes to form carbon monoxide (CO) and sulfur (S) in the stratosphere.

carcinogenic (carcinogen) Refers to a substance capable of causing cancer. See encyclopedia entry.

carnivore An organism that eats meat.

carrying capacity The maximum number of people, or individuals of a particular species, that a given part of the environment can maintain indefinitely. See encyclopedia entry.

catalytic converter An air pollution control device used in the exhaust system of cars. It helps complete combustion of any fuel that was not burned in the engine and changes the unburned hydrocarbons and carbon monoxide in the exhaust into carbon dioxide and water vapor.

census An official count of population.

chlorofluorocarbons (CFCs) A family of chemicals commonly used in air conditioners and refrigerators as coolants and as solvents and aerosol propellants. CFCs drift into the upper atmosphere where their chlorine components destroy ozone. See encyclopedia entry "CFC".

class In taxonomy, a category just beneath the phylum and above the order; a group of related, similar orders.

clear-cutting Harvesting all the trees in one area at one time, a practice that eliminates local habitat and biodiversity. See encyclopedia entry.

climate change Associated with global warming and the greenhouse effect, climate change frequently refers to the buildup in the atmosphere of artificial gases that trap the sun's heat, causing changes in weather patterns on a global scale. The effects include changes in rainfall patterns, increases in sea level, potential droughts, habitat loss, and heat stress. See encyclopedia entry.

cloning In biotechnology, obtaining a group of genetically identical cells from a single cell; making identical copies of a gene. Clones are individuals all derived asexually from the same single parent. See encyclopedia entry.

closure The procedure an operator must go through when a landfill reaches the legal capacity for solid waste. No more waste can be accepted and a cap is usually placed over the site. Post-closure care includes monitoring groundwater, landfill gases, and leachate collection systems.

cogeneration The process of producing both electricity and thermal energy, such as heat or steam, which can then be used for industrial and commercial heating or cooling. See encyclopedia entry.

co-management The sharing of authority, responsibility, and benefits between government and local communities.

commercial waste Solid waste from businesses.

commercial waste management facility A treatment, storage, disposal, or transfer facility that accepts wastes for profit.

community An integrated group of species inhabiting a given area; the organisms within a community influence one another's distribution, abundance, and evolution. See encyclopedia entry.

compost Decayed organic matter that can be used as a fertilizer or soil additive. See encyclopedia entry "Composting."

condensation Gaseous water vapor that begins to change to tiny water droplets or ice crystals when the air gets cold enough.

cone of depression A lowering in the water table that develops around a pumped well.

consent decree A legal document submitted by the Department of Justice on behalf of the Environmental Protection Agency for approval by a federal judge to settle a case. A consent decree can be used to formalize an agreement reached between EPA and potentially responsible parties (PRPs) for cleanup at a Superfund site. They are also signed by regulated facilities to cease or correct certain actions or processes that are polluting the environment and include payment of penalties.

conservation Preserving and renewing natural resources to assure their highest economic or social benefit over the longest period of time. See encyclopedia entry, "Conservation Movement, U.S."

consumer Any organism that must consume other organisms to satisfy its energy needs.

continental drift A term applied to early theories supporting the possibility that the continents are in motion over the earth's surface.

continuous discharge A permitted release of pollutants into the environment that occurs without interruption, except for infrequent shutdowns for maintenance, process changes, and so forth.

controlled reaction A chemical reaction at temperature and pressure conditions that are maintained within safe limits to produce a desired product.

convection Process by which, in a fluid being heated, the warmer part of the mass will rise and the cooler portions sink.

coriolis force A force resulting from the earth's rotation affecting the oceans and atmosphere, causing hurricanes and whirlpools to rotate counterclockwise in the Northern Hemisphere and clockwise in the Southern Hemisphere.

corrosive Refers to a substance that gradually eats or wears away materials by chemical action.

county emergency operations plan A plan required by Federal Emergency Management Agency regulations that describes actions the county will take to respond to emergency situations such as natural disasters, major fires, transportation incidents, or chemical releases.

cryosphere The frozen part of the earth's surface.

crystallization Physical or chemical process or action that results in the formation of solid forms that are regularly shaped, sized, and patterned.

cubic feet per second (CFS) Standard unit for measurement of stream flow or wastewater discharge.

cultivar A cultivated variety (genetic strain) of a domesticated crop plant.

cumulus cloud Type of cloud in the troposphere that is vertically shaped with flat bases and fluffy, rounded tops because of buoyant upward convection during warm summer weather.

current A flow of electrons in an electrical conductor.

cycle In the atmosphere or biosphere, a sequence of events in repetitive motion in which the final output feeds back into the initial input.

dechlorination Removal of chlorine and its chemical replacement with hydrogen or hydroxide ions to detoxify a substance.

decommissioning The act of retiring or dismantling a dam, power plant, or other form of equipment.

decomposition The breakdown of organic materials by organisms in the environment, releasing energy and simple organic and inorganic compounds. Consumers break down organic material (such as sugars and proteins) to obtain energy for their own growth.

deep well injection A process by which waste fluids are injected deep below the surface of the earth. See encyclopedia entry.

deforestation Processes that result in the change of forested lands to nonforest uses. See encyclopedia entry.

delist Use of the petition process (1) to have a chemical's toxic designation rescinded; (2) to remove a site from the National Priority List; or (3) to exclude a particular waste from regulation even though it is a listed hazardous waste.

delta Fan-shaped mass of sediments deposited where a river discharges to a larger, slower moving body of water.

demography The rate of growth and age structure of populations; also the processes that determine these properties. See encyclopedia entry.

denitrification Natural chemical conversion of dissolved nitrogen (nitrite) to gaseous nitrogen.

desertification Destruction of an existing vegetative area to form desert. See encyclopedia entry.

detritus Decaying organic material.

discharge The release of any waste into the environment from a point source. Usually refers to the release of a liquid waste into a body of water through an outlet such as a pipe, but also refers to air emissions.

discharge zone An area of land where there is a net annual transfer of water from the groundwater to surface water, such as to streams, springs, lakes, and wetlands.

disease When normal function is impaired in an organism by a genetic disorder or from the activity of a parasite or other organism living within it, such as viruses, bacteria, or fungi.

dispersal Spread of a species or substance to a new location.

dispersion model A mathematical prediction of how pollutants from a discharge or emission source will be distributed in the surrounding environment under given conditions of wind, temperature, humidity, and other environmental factors.

disposal facility A landfill, incinerator, or other facility that receives waste for disposal. The facility may have one or many disposal methods available for use.

dissolved oxygen Oxygen that is freely available in water to sustain the lives of fish and other aquatic organisms. See encyclopedia entry.

diurnal tide Tide occurring once a day.

easement A limited right of ownership of one's land conveyed by deed to another for a special purpose.

ecological niche Where a species lives, what it consumes, and how it avoids consumption by predators or displacement by other species.

ecology The study of the relationships between all living organisms and the environment, especially the pattern of interactions. See encyclopedia entry "Ecology as a Science."

ecosystem The interaction of all living organisms in a particular environment; every plant, insect, aquatic animal, bird, or land species that forms a complex web of interdependency. See encyclopedia entry.

ecotone A border between two biomes, where the plants and animals of those biomes mingle.

ecotourism Travel undertaken to witness sites or regions of unique natural or ecologic quality, or the provision of services to facilitate such travel.

effluent Wastewater discharged from a point source, such as a pipe.

El Niño Abnormally warm seasonal changes occurring every 3 to 5 years. See encyclopedia entry.

emission The release or discharge of gases or particles into the air.

emission standards Government standards that establish limits on discharges of pollutants into the environment.

endangered species Animals, plants, birds, fish, or other living organisms threatened with extinction by artificial or natural changes in the environment. See encyclopedia entry.

energy efficiency Programs aimed at reducing the energy used by specific devices and systems without affecting the services provided. See encyclopedia entry.

energy recovery To capture energy from waste through any of a variety of processes, such as burning. Many new technology incinerators are waste-to-energy recovery units.

Environmental Assessment (EA) A preliminary environmental analysis required by the National Environmental Policy Act to determine whether a federal activity such as building airports or highways would significantly affect the environment. See encyclopedia entry.

Environmental Impact Statement (EIS) A document prepared by or for the Environmental Protection Agency that identifies and analyzes environmental impacts of a proposed action. The EIS describes positive and negative effects and lists alternative actions.

environmental response team (ERT) Environmental Protection Agency's group of highly trained scientists and engineers whose capabilities include multimedia sampling and analysis,

hazard assessment, hazardous substance and oil spill cleanup techniques, and technical support.

epidemiologist A medical scientist who studies the various factors involved in the incidence, distribution, and control of disease in a population. See encyclopedia entry "Epidemiology".

erosion The wearing away of soil by wind or water, intensified by land-clearing practices related to farming, residential or industrial development, road building, or logging. See encyclopedia entry "Soil erosion."

estuary A complex ecosystem along a coastline where freshwaters mix with saltwater, such as a bay or lagoon.

eutrophication A process of overenrichment of a body of water with nutrients, resulting in overgrowth of algae, algae die-offs, and oxygen depletion. See encyclopedia entry.

evolution Any gradual change. Organic evolution is any genetic change in organisms from generation to generation. See encyclopedia entry.

exotic species A species occurring in an area outside its natural range. See encyclopedia entry.

extinction The evolutionary termination of a species caused by the failure to reproduce and the death of all remaining members of the species; the natural failure to adapt to environmental change.

fauna All of the animals found in a given area.

fecal coliform bacteria Found in the intestinal tracts of mammals, this bacteria in water or sludge is an indicator of pollution and possible contamination by pathogens.

Federal Emergency Management Agency (FEMA) A federal government agency having responsibilities in hazard relief.

feedback cycle Increase of the input into a system, which produces a negative or a positive output; the effect the output plays on the input.

feedback mechanism A mechanism that connects one aspect of a system to another, either amplifying or moderating the effect.

feedstock Raw material supplied to a machine or processing plant from which other products can be made.

fertilization Efforts to enhance plant growth by using nitrogen-based fertilizer or increased deposition of nitrates in precipitation. See encyclopedia entry "Fertilizer."

fish ladder A series of ascending pools of running water constructed to enable fish to swim upstream and around or over a dam.

flash point The lowest temperature at which evaporation of a substance produces enough vapor to form an ignitable (flammable) mixture with air.

floodplain Mostly level land along rivers and streams that may be submerged by floodwater. See encyclopedia entry.

flora All of the plants found in a given area.

fluorocarbons Carbon-fluorine compounds that often contain other elements such as hydrogen, chlorine, or bromine.

food chain A process of energy transfer from green plants to grazers and predators. See encyclopedia entry.

food web A system of interlocking food chains.

fossil fuel Natural resources that contain stored energy from the sun, which is released upon combustion. See encyclopedia entry "Fuel, Fossil."

Framework Convention on Climate Change (FCCC) International treaty committing signatory countries to stabilize human-induced greenhouse gas emissions to levels that are not dangerous to the climate system.

free radicals Highly reactive atoms or molecules with an uneven number of electrons.

freon Stable liquids or gases usually produced for solvents, aerosol propellants, refrigeration, air conditioning, or styrofoam.

frugivore An animal that primarily eats fruit.

fry The transitional stage of recently hatched fish that spans from absorption of the yolk sac through several weeks of independent feeding.

fugitive emissions Air pollutants released to the air other than those from stacks or vents; typically small releases from leaks in plant equipment.

fungicide A pesticide used to control or destroy fungi.

Gaia hypothesis A hypothesis proposed during the early 1970s by James Lovelock stating that all living organisms have the ability to affect their surroundings, such as atmosphere and climate, to maximize biological success. See encyclopedia entry.

gas turbine plant A plant in which the prime mover is a gas turbine. A gas turbine consists typically of an axial-flow air compressor and one or more combustion chambers, where liquid or gaseous fuel is burned. The hot gases are passed to the turbine and expand to drive the generator and are then used to run the compressor.

gene The functional unit of heredity; the part of the DNA molecule that encodes single enzyme or structural protein units. See encyclopedia entry "Genetics".

gene bank A facility established for the conservation of individuals (e.g., seeds), of tissues, or of reproductive cells of plants or animals.

general circulation model (GCM) A global, three-dimensional computer model of the climate system that can be used to simulate human-induced climate change.

generating unit Any combination of physically connected generators or other prime movers operated together to produce electric power.

genetic engineering A process of inserting new genetic information into existing cells in order to modify an organism for the purpose of changing particular characteristics.

genotype The set of genes possessed by an individual organism.

geosphere The soils, sediments, and rock layers of the earth's crust, both continental and beneath the ocean floors.

gigawatt (GW) One billion watts.

gigawatthour (GWh) One billion watthours.

glaciation The process of forming glaciers.

glacier A slow-moving mass of ice formed in higher latitudes and elevations. See encyclopedia entry.

global warming The overall increase of atmospheric temperature due to a buildup of greenhouse gases. See encyclopedia entry.

Gondwanaland A hypothetical supercontinent including the continents of the Southern Hemisphere.

grassroots Refers to organizations or movements involving people or society at a local level rather than at the center of major political activity.

greenhouse effect The act in which outgoing infrared radiation that would normally exit from a planet's atmosphere is trapped or reflected. See encyclopedia entry "Greenhouse Effect."

greenhouse gases Atmospheric components that absorb radiation in the infrared region of the spectrum. Infrared radiation is reflected and emitted by the earth's surface as heat, causing a fairly large warming effect when trapped by these gases in the atmosphere.

grid The layout of an electrical distribution system.

groundwater Water found below the surface of the land, usually in porous rock formations. Groundwater is the source of water found in wells and springs and is used for drinking. See encyclopedia entry.

ha Abbreviation of *hectare*, a unit of area measurement in the metric system equal to 10,000 square meters or 2.47 acres.

habitat The environment in which an organism lives.

health assessment An evaluation of available data on existing or potential health risks.

heavy metal A common hazardous waste; can damage organisms at low concentrations and tends to accumulate in the food chain.

herbicide A pesticide designed to control or kill plants, weeds, or grasses. These chemicals may have wide-ranging effects on nontarget species other than those the pesticide is meant to control. See encyclopedia entry.

herbivore An organism that eats plants.

heterotrophic A species or entire ecosystem that acquires its energy by import rather than producing its own food.

host An organism that serves as the habitat for a parasite.

household or **domestic waste** Solid waste from households or apartment buildings. Domestic waste may contain a significant amount of toxic or hazardous waste from improperly discarded pesticides, paints, batteries, and cleaners.

hybridization Crossing of individuals from genetically different strains, populations, or species.

hydrocarbons Chemicals that consist entirely of hydrogen and carbon. Hydrocarbons contribute to air pollution problems like smog.

hydrochlorofluorocarbons (HCFCs) A useful replacement for CFCs, which have the ability to destroy ozone; HCFCs, however, are chemically unstable.

hydroelectric plant A plant in which the turbine generators are driven by falling water.

hydrogen chloride (HCl) A colorless, corrosive, nonflammable gas.

hydrogen peroxide (H_2O_2) A colorless, unstable oxidant with a bitter taste and caustic to the skin; decomposes, liberating oxygen.

hydrogen sulfide (H_2S) A relatively unstable compound that nevertheless survives for a long time because of its slow reaction with atmospheric oxygen.

hydrologic cycle Continuous cycle of water in the biosphere.

hydroperoxy radical (HO_2) A radical that readily reacts with nitrogen oxides and hydrocarbons in the atmosphere; its major source is formaldehyde photolysis.

hydrosphere The part of the atmosphere that contains water in the liquid, solid, or gaseous phase.

hydroxyl radical (OH) Does not normally exist in a stable form; readily reacts with methane and carbon monoxide.

hypoxia Depletion of dissolved oxygen in water to low levels due to oxygen-demanding compounds or nutrient overenrichment.

ice core A sampling of ice used to determine the historic chemical composition of the atmosphere.

identification code (EPA I.D. number) The unique code assigned to each generator, transporter, and treatment, storage, or disposal facility by the Environmental Protection Agency to facilitate identification and tracking of hazardous waste. Superfund sites also have assigned identification numbers.

impoundment A body of water or sludge confined by a dam, dike, floodgate, or other barrier.

incineration The destruction of solid, liquid, or gaseous wastes by controlled burning at high temperatures. Hazardous organic compounds are converted to ash, carbon dioxide, and water. See encyclopedia entry.

indicator species A species whose status provides information on the overall condition of the ecosystem and of other species in that ecosystem.

indigenous peoples People whose ancestors inhabited a place or country when persons from another culture or ethnic background arrived on the scene and frequently dominated them through conquest, settlement, or other means. Indigenous peoples often live more in conformity with their own social, economic, and cultural customs and traditions than according to those of the dominant culture. See encyclopedia entry.

industrial waste Unwanted materials produced in or eliminated from an industrial operation and categorized as liquid waste, sludge, solid waste, and hazardous waste.

inert ingredients Substances that are not active, such as water and petroleum distillates. In pesticides, inert ingredients do not attack a particular pest, but some are chemically or biologically active, causing health and environmental problems.

infrared radiation (IR) Energy that is emitted in the form of electromagnetic waves. The emission of these wavelengths from the earth's surface is what contributes to the process that heats the atmosphere.

insecticide A pesticide compound specifically used to kill or prevent the growth of insects.

integrated pest management (IPM) A combination of biological, cultural, and genetic pest control methods with use of pesticides as the last resort. See encyclopedia entry "Pest Control."

Intergovernmental Panel on Climate Change (IPCC) Group of leading experts on climate change and environmental, social, and economic sciences from over 60 nations who help prepare periodic assessments of the scientific underpinnings for understanding global climate change and its consequences.

internal combustion plant A plant in which the prime mover is an internal combustion engine, which has one or more cylinders converting energy released from the rapid burning of a fuel-air mixture into mechanical energy.

interstate commerce A clause in the U.S. Constitution that reserves to the federal government the right to regulate the conduct of business across state lines.

introduced species A species occurring in an area outside its historically known natural range as a result of intentional or accidental dispersal by human activities.

inversion An atmospheric condition caused by increasing temperature with elevation, resulting in a layer of warm air preventing the rise of cooler air trapped beneath. This condition prevents the rise of pollutants that might otherwise be dispersed, increasing ozone to harmful levels.

irradiated food Food that has been briefly exposed to radioactivity (usually gamma rays) to kill insects, bacteria, and mold. See encyclopedia entry "Irradiation, Food."

irrigation The controlled application of water to arable lands to supply water requirements not satisfied by rainfall. See encyclopedia entry.

isotope Two or more forms of an element that have the same atomic number but different numbers of neutrons in the nucleus and, therefore, a different atomic weight.

joule (J) An international unit of measure that expresses a unit of energy equal to the work done by a force of one newton acting through one meter.

keystone species A species whose loss from an ecosystem would cause a great change in other species populations or in ecosystem processes. See encyclopedia entry.

kilowatt (kW) One thousand watts.
kilowatthour (kWh) One thousand watthours.
Kyoto Protocol Unratified agreement made in 1997 committing industrialized nations to cut greenhouse gas emissions.

lacustrine Relating to a lake environment.
lagoon A shallow, artificial treatment pond where sunlight, bacterial action, and oxygen work to purify wastewater; a stabilization pond.
LANDSAT A robotic satellite system that takes pictures of the earth's surface and digitally transmits them to earth.
leachate Liquid (mainly water) that percolates through a landfill and has picked up dissolved, suspended, or microbial contaminants from the waste.
lethal dose 50 (LD 50) The dose of a toxicant that will kill 50 percent of test organisms within a designated period of time. The lower the LD 50, the more toxic the compound.
levee A long, narrow embankment made of earth built to protect land from flooding.
limnology The study of river system ecology and life.
liner Structure of natural clay or manufactured material (plastic) that serves as a barrier to restrict leachate from reaching or mixing with groundwater in landfills, lagoons, or other impoundments.
lithosphere The outer, rigid shell of the earth containing the crust, continents, and plates.
littoral zone The area on or near the shore of a body of water.
load (electric) The amount of electric power delivered or required at any specific point or points in a system.

maximum contaminant level (MCL) The maximum level of certain contaminants permitted in drinking water supplied by a public water system as set by the Environmental Protection Agency under the federal Safe Drinking Water Act.
mesopause The transition zone between the mesosphere and the thermosphere; the coldest portion of the mesosphere.

mesosphere In the atmosphere, the region above the stratosphere and below the thermosphere.

meteorology The science of the atmosphere and its direct effects on the earth's surface. See encyclopedia entry.

methane (CH$_4$) A colorless, odorless, flammable, greenhouse gas; it is released naturally into the air from anaerobic environments such as marshes, swamps, and rice fields.

metric ton Common international unit of measure used to indicate emissions. A metric ton is equal to 2,205 pounds.

midocean ridge A major elevated linear feature of the seafloor consisting of many small, slightly offset segments; occurs where two plates are being pulled apart and new oceanic lithosphere is being created.

mitigation Measures taken to compensate for damage to natural systems caused by a particular project or human activity.

monitoring well A well used to take water quality samples or to measure groundwater levels.

Montreal Protocol An international environmental agreement to prevent the use of substances that are harmful to the ozone layer.

morbidity Rate of incidence of disease.

mortality Death rate.

national ambient air quality standards (NAAQS) Maximum air pollutant standards set by the Environmental Protection Agency under the authority of the Clean Air Act.

National Center for Atmospheric Research (NCAR) A nonprofit organization dedicated to furthering understanding of the earth's atmosphere.

National Oceanic and Atmospheric Administration (NOAA) A U.S. government agency that determines how oceans and atmosphere should be developed, regulated, and analyzed.

National Response Center (NRC) The primary communications center operated by the U.S. Coast Guard to receive reports of major chemical and oil spills and other hazardous conditions.

National Response Team (NRT) Representatives from fifteen federal agencies with interests and expertise in various aspects of emergency response to incidents of pollution.

native species Plants, animals, fungi, and microorganisms that occur naturally in a given area or region.

natural gas Naturally occurring mixture of hydrocarbon and nonhydrocarbon gases found in porous geological formations beneath the earth's surface, often in association with petroleum. See encyclopedia entry.

nekton Aquatic animals, such as fish, with enough power to overcome currents.

neutralization The chemical process in which the acidic or basic characteristics of a fluid are changed to those of water.

niche The portion of the environment occupied by a species.

nitrification The conversion of organic nitrogen compounds to inorganic nitrogen compounds by bacteria.

nitrogen (N$_2$) A colorless, tasteless, odorless gas that makes up 78 percent of the atmosphere. See encyclopedia entry.

nitrogen fixation A process whereby nitrogen-fixing bacteria that live with plants convert atmospheric nitrogen to nitrogen compounds that plants can use directly.

nitrogen monoxide (NO) Compound that can act as a catalyst in the reactions that cause the destruction of ozone.

nitrous oxide (N$_2$O) A principal greenhouse gas that absorbs infrared radiation.

nocturnal Refers to activity only at night.

nongovernmental organization (NGO) A nonprofit group or association organized outside institutionalized political structures to realize particular social objectives (such as environmental protection) or serve particular constituencies(such as indigenous peoples). Activities include research, training, local organization, legal advocacy, lobbying for legislative change, and civil disobedience.

nonpoint source Any source of pollution not associated with a distinct discharge point. Examples include rainwater; runoff from agricultural

lands, industrial sites, parking lots, and timber operations; escaping gases from pipes and fittings.

ohm Unit of measure of electrical resistance. The resistance of a circuit in which a potential difference of 1 volt produces a current of 1 amp.

omnivore An organism that eats both plants and animals.

opportunistic species Species able to quickly adapt to environmental changes.

organically grown Food, feed crops, and livestock grown within an intentionally diversified, self-sustaining agro-ecosystem.

outgassing The release of gases into the earth's atmosphere by volcanic activity.

oxidant A substance containing oxygen that reacts chemically with other materials to produce new substances. Oxidants are the primary ingredients in smog.

ozone hole The total amount of ozone in the atmosphere over Antarctica that was discovered to be decreasing.

ozone layer The layer of the stratosphere that absorbs UV radiation, creating a warm layer of air. See encyclopedia entry.

Pangaea A huge protocontinent thought to have existed 200 million years ago and from which all present continents are thought to have formed.

parasite An organism that lives on or within another organism on which it feeds. See encyclopedia entry "Parasitism."

particulate Liquid or solid particles such as dust, smoke, mist, or smog found in air emissions.

parts per million (ppm) and **parts per billion (ppb)** Terms of measurement of small amounts of a substance contained in a sample.

patent A government grant of temporary monopoly rights on innovative processes or products.

pathogen A disease-causing microbe.

PCV system An air pollution control device in automobile engines.

peak demand The maximum load during a specified period of time.

pelagic Refers to organisms living in open waters.

permafrost Soil, silt, and rock located in perpetually cold areas that remain frozen year-round.

permeability The ease with which water or other fluids pass through a substance.

permit A legal document issued by state or federal authorities containing a detailed description of the proposed activity, operating procedures, requirements, and regulations.

pesticide A substance intended to repel, kill, or control any species designated a pest. Pesticides include herbicides, insecticides, rodenticides, fungicides, and bactericides. See encyclopedia entry.

petroleum A mixture of hydrocarbons existing in the liquid state found in natural underground reservoirs, often associated with natural gas. See encyclopedia entry.

pH The measure of acidity or alkalinity of a chemical solution from 0–14. Anything neutral, for example, has a pH of 7. Acids have a pH less than 7, bases (alkaline substances) greater than 7.

phenol A corrosive, poisonous, crystalline acidic compound found in coal tar and wood tar; phenols are released into the atmosphere when coal and wood are burned.

phenotype The properties of an organism that develop through the interaction of genes and environment.

photic zone Upper portion of the water, admitting sufficient light for photosynthesis.

photolysis Chemical decomposition brought about through the action of energy in the form of electromagnetic waves.

photosynthesis A biological process that plays a vital role in cycling the atmosphere's carbon dioxide while producing oxygen and other complex substances. See encyclopedia entry.

phylogenetic Pertaining to the evolutionary history of a particular group of organisms.

phylum A high-level category of organisms just beneath the kingdom and above the class; a group of related, similar classes.

phytoplankton Green plants, such as algae, inhabiting waters, unattached and drifting with the currents.

plankton Forms of marine organic life that gather vital energy for life through photosynthesis. See encyclopedia entry.

plate tectonics Process of planetary surface plate formation, movement, interaction, and destruction that causes earthquakes.

plume A concentration of contaminants in air, soil, or water, usually extending from a distinct source.

point source A stationary location or fixed facility such as an industry or municipality that discharges pollutants into air or surface water through pipes, ditches, lagoons, wells, or stacks.

point source pollution Pollution into bodies of water from specific discharge points such as sewer outfalls or industrial waste pipes.

pollution Any substances in water, soil, or air that degrade the natural quality of the environment, offend the senses of sight, taste, or smell, or cause a health hazard.

polychlorinated biphenyls (PCBs) A group of toxic chemicals used for insulation in electrical transformers and capacitors, and as a lubricant in gas pipeline systems. See encyclopedia entry "PCB."

population Organisms of a given species.

potable water Raw or treated water that is considered safe to drink.

Precambrian All geologic time from the beginning of earth history to 570 million years ago. Also refers to the rocks formed in that epoch.

precipitation Snow, rain, sleet, or hail.

pretreatment Methods used by industry sources of wastewater to remove, reduce, or alter the pollutants in wastewater before discharge to a treatment facility.

primary consumer An organism deriving its energy directly from green plants.

primary producer Green plants capable of photosynthesis; the base of the food chain.

process wastewater Any water that comes into contact with any raw material, product, by-product, or waste.

producer Any organism that brings energy into an ecosystem from inorganic sources.

protected area A legally established land or water area that is regulated and managed to achieve specific conservation objectives.

pumped-storage hydroelectric plant A plant that generates electric energy during peak load periods by using water previously pumped into an elevated storage reservoir during off-peak periods when excess generating capacity is available.

quality assurance/quality control A system of procedures, checks, audits, and corrective actions to ensure that all technical, operational, monitoring, and reporting activities are of the highest quality.

radical A highly reactive molecule or atom with an unpaired electron.

radon A colorless, naturally occurring gas formed by radioactive decay of radium atoms. See encyclopedia entry.

reach Distance between two specific points outlining a portion of a stream or river.

reactivity Refers to those hazardous wastes that are normally unstable and readily undergo violent chemical change but do not explode.

recharge zone An area where rainwater soaks through the earth to reach an aquifer.

recycling Reusing materials and objects in original or changed forms rather than discarding them as wastes. See encyclopedia entry.

red tide An algae bloom that manufactures biotoxins, causing fish kills and poisoning in human consumers.

registration Formal listing with the Environmental Protection Agency of a new pesticide before sale or distribution.

rehabilitation The recovery of specific ecosystem services in a degraded ecosystem or habitat.

release Any spilling, leaking, pumping, pouring, emitting, dumping, or disposing into the environment of a hazardous or toxic chemical or extremely hazardous substance.

reservoir An artificial facility, often behind a dam or other obstruction, for the storage, regulation, and controlled release of water.

resource recovery The extraction of useful materials or energy from solid waste. Such materials can include paper, glass, and metals that can be reprocessed for re-use.

restoration The return of an ecosystem or habitat to its original community structure, natural group of species, and natural functions.

riparian zone The habitat found on stream banks and river banks where semi-aquatic and terrestrial organisms mingle.

risk A measure of the chance that damage to life, health, property, or the environment will occur. See encyclopedia entry.

risk analysis The estimation of hazards associated with contaminants or other environmental conditions as they affect exposed humans or the ecosystem. See encyclopedia entry "Risk-Benefit Analysis."

risk assessment A process to determine the increased risk from exposure to environmental pollutants together with an estimate of the severity of impact. Risk assessments use specific chemical information plus risk factors.

risk factor A characteristic (e.g., sex, age, and race) or variable (e.g., degree of exposure) associated with increased chance of toxic effects.

salinity Salt concentration in marine waters.

saprophyte An organism that feeds on dead and decaying organisms.

scavenger An animal that eats dead remains and wastes of plants and other animals.

scrubbing Removal of impurities by spraying a liquid that concentrates the impurities into waste.

sea level The average level of the ocean over the entire earth.

sedentary Not moving; organisms, both plants and animals, that spend the majority of their lives in one place.

sediment Topsoil, sand, and minerals washed from the land into water, usually after rain or snow melt. A collection in rivers, reservoirs, and harbors can destroy fish and wildlife habitats.

seedbank A facility designed for the conservation of individual plant varieties through seed preservation and storage.

sentinel species A species that can provide advanced warning of more generalized environmental degradation.

short ton A unit of weight equal to 2,000 pounds.

silt Sediment composed of particles between .004 mm and .06 mm in diameter.

silviculture Use and management of forest resources.

sink A natural removal process through rain of pollutants in the troposphere.

siting Choosing a location for a facility.

SI unit A unit of measure in the International System, or Système internationale d'unités; a modernized and expanded version of the metric system.

sludge The residue (solids and some water) produced as a result of raw or wastewater treatment.

slurry A pumpable mixture of solids and fluid.

smog Dust, smoke, or chemical fumes that pollute the air and make hazy, unhealthy conditions. See encyclopedia entry.

smoke The product of incomplete combustion, consisting of carbon and small liquid and solid particles.

smolt A young salmon or steelhead migrating to the ocean and undergoing physiological changes to adapt its body from a freshwater to a saltwater environment.

solid waste Any solid or semi-solid liquid or contained gaseous material discarded from industrial, commercial, mining, or agricultural operations, or community activities. See encyclopedia entry "Waste Management."

solid waste management facility Any disposal or resource recovery system; any system, program, or facility for resource conservation; any facility for the treatment of solid wastes.

sorption A class of processes by which one material is taken up by another. Absorption is the process of the penetration of one material into another; adsorption is the action of one material being collected on another's surface.

source reduction The design, manufacture, purchase, or use of materials (such as products and packaging) to reduce the amount or toxicity of garbage generated.

spawning The releasing and fertilizing of eggs by fish.

specialist Organism that has adopted a lifestyle specific to a particular set of conditions.

species A group of organisms capable of interbreeding freely with each other but not with members of other species.

spillway A structure over or through which excess waters are discharged.

stakeholder An individual or organization with a substantive connection to the use of a natural resource.

steam-electric plant (conventional) A plant in which the prime mover is a steam turbine.

stratification Vertical separation of water masses into layers with different characteristics.

stratosphere The thermal atmospheric region between the troposphere and the mesosphere.

stratospheric ozone Also called the ozone layer; prevents most of the sun's ultraviolet solar radiation from reaching the earth's surface.

subspecies A subdivision of a species; a population or series of populations differing genetically from other subspecies of the same species.

succession Predictable changes in communities following a natural or human disturbance. For example, after a gap is made in a forest by logging, clearing, fire, or treefall, the first species to re-emerge are often shade-intolerant and are eventually replaced by shade-tolerant species that can grow beneath the initial species.

surface water All water naturally open to the atmosphere (rivers, lakes, reservoirs, ponds, streams, seas, estuaries) and all springs, wells, or other collectors directly influenced by surface water.

suspension feeder An organism that feeds on plankton.

sustainable agriculture Environmentally friendly methods of farming that allow the production of crops or livestock without damage to the farm as an ecosystem, including effects on soil, water supplies, or other surrounding natural resources. See encyclopedia entry.

sustainable development Development that meets the needs and aspirations of the current generation without compromising the ability to meet those of future generations.

switching station Facility equipment used to tie together two or more electric circuits through switches. The switches are selectively arranged to permit a circuit to be disconnected or to change the electric connection between the circuits.

symbiosis Living together. Three types of relationships can be identified: mutual, in which organisms rely on each other for survival; parasitism, where one organism benefits at the other's expense; and commensalism, where one partner benefits and the other is neither benefited nor harmed. See encyclopedia entry.

synergism Two or more organisms working together to produce a greater total result than the sum of their independent effects.

taxon The classification unit individuals or sets of species are assigned, such as Homo sapiens.

taxonomy The naming and assignment of organisms to particular groups.

tectonics The structural arrangement of rocks in any crust of a planet.

temperate Refers to seasonal change in temperature and moisture.

ten-to-the-minus-sixth (10_{-6}) Used in risk assessments to refer to the probability of risk. Literally means a chance of one in a million. Similarly, ten-to-the-minus-fifth means a probability of one in 100,000.

terrestrial Living on land.

thermal pollution Ordinarily, the human-caused change in water temperature that results in damage to aquatic life. See encyclopedia entry.

thermodynamics The science of heat and temperature; laws governing the conversion of heat into mechanical, electrical, or chemical energy.

tolerance Permissible residue level for pesticides in raw agricultural produce and processed foods.

toxic substance A chemical or mixture that can cause illness, death, disease, or birth defects. Many toxic substances are pollutants and contaminants in the environment.

trace gases Gases in the atmosphere that do not occur in large quantities but are significant to life on earth.

transformer An electrical device for changing the voltage of alternating current.

trophic level Position in the food chain, determined by the number of energy-transfer steps to that level.

tropical Refers to regions in which the climate undergoes little seasonal change in either temperature or rainfall.

troposphere The lowest layer of the earth's atmosphere, ranging from the ground to the stratosphere.

turbidity Cloudiness of water, caused by suspended material or plankton.

turbine A machine for generating rotary mechanical power from the energy of a stream of fluid, such as water, steam, or hot gas.

ultraviolet ray Radiation from the sun in the invisible portion of the spectrum. Some UV rays (UV-A) enhance plant life and are useful in certain medical and dental procedures. Other UV rays (UV-B) can cause skin cancer or other tissue damage.

upwelling The raising of benthic nutrients to the surface waters.

UV radiation Energy that is emitted in the form of electromagnetic waves composed of UV-A, UV-B, and UV-C light.

UV-B An ultraviolet wavelength of light whose intensity increases at the earth's surface because of decreases in stratospheric ozone.

vadose zone The unsaturated zone lying between the earth's surface and the top of groundwater.

valence electrons The outermost electrons of an atom.

vapor The gas given off by substances that are solids or liquids at ordinary atmospheric pressure and temperatures.

vapor dispersion The movement of vapor clouds or plumes in the air due to wind, gravity, spreading, and mixing.

vent The connection and piping through which gases enter and exit a piece of equipment.

vertical mixing The movement of air in a vertical direction, usually caused by differences in temperature and density.

wastewater treatment plant A facility containing a series of tanks, screens, filters, and other processes by which pollutants are removed from water. Most treatments include chlorination to attain safe drinking water standards.

water column The portion of an aquatic or marine environment extending from the water surface to the bottom.

water cycle The process by which water is transpired and evaporated from the land and water, condensed in the clouds, and precipitated onto the earth again to replenish water in the bodies of water on the earth. See encyclopedia entry.

water quality The condition of water as determined by measurements of suspended solids, acidity, turbidity, dissolved oxygen, and temperature and by the presence of organic matter and pollution chemicals.

water quality standards (WQS) The maximum concentration of a pollutant permitted in a body of water to protect its original use.

water rights A legal right to use a specific amount of water from a natural or artificial body of water for irrigation, mining, power or domestic use.

watershed An area of land drained by a river or stream. See encyclopedia entry.

water table The boundary between the saturated and unsaturated zones of the earth. Generally, the level to which water will rise in a well (except artesian wells).

watthour (Wh) An electrical energy unit of measure equal to 1 watt of power supplied to, or taken from, an electric circuit steadily for 1 hour.

wetland An area where saturation with water is the dominant influence on characteristics of the soil and plant community. See encyclopedia entry.

wildlife refuge An area designated for the protection of wild animals, within which hunting and fishing are either prohibited or strictly controlled.

xenobiotic Refers to artificial substances found in the environment (e.g., plastics).

zone of saturation The layer beneath the surface of the land in which all openings are filled with water.

zooplankton Tiny, free-floating organisms in aquatic systems.

Resources for Further Study

Prepared by Frederick W. Stoss, Science
and Engineering Library,
University at Buffalo, New York

Using Your Libraries

Professor Marta Dosa, now Professor Emerita in the School of Information Studies at Syracuse University, provided a definition of environmental information as "the process that transfers data and information from its source to users in any field of knowledge or activity applicable to environmental problem solving." Environmental information is needed by students, teachers, and members of the community to understand the complex and dynamic natures of environmental problems, to challenge conventional wisdom and stimulate new ideas, to discuss issues and make decisions, and to respond to social and political conditions.

    Libraries have been around for a long time as archives for ideas and knowledge and as institutions that preserve and provide access to this information. Libraries are designed and managed to provide the means by which this information is stored (e.g., collections of books, reports, periodicals audiovisual resources, electronic and digital databases), services (reference and referral, education and instruction, current awareness), and information products (pathfinders and resource guides, bibliographies, pamphlets and brochures, guides to the literature). Libraries are often the most trusted, used, and respected of all community institutions. There are a variety of libraries at your disposal.

School libraries and media centers provide resources for teachers and students to teach and learn. The collections of school libraries focus on books, magazines, and audiovisual resources published or produced at appropriate grade levels. School librarians often provide instruction on the use of libraries and work with teachers in providing class-specific exercises, projects, and activities. Many school libraries provide access to the Internet, which opens new worlds of information, including other libraries.

Public libraries have been an integral part of the education system and process for decades. The collections of public libraries vary from one community to another. Small rural communities may have a small library with limited resources. Their collections reflect the needs of the communities they serve and are often rich sources for local information. In larger cities the public library is provided not as a single library, but as a network of neighborhood branch libraries with larger more comprehensive collections. County library systems may link town and city libraries in a more comprehensive network. The collections of larger public libraries serve as the corporate library or technical information center for the area's small business community, local government agencies, and nonprofit groups. They have more technical and discipline-oriented collections of reports, books, government documents, magazines, technical and professional journals, as well as other audiovisual resources. Many public libraries and library systems supplement the collections of the school library and are used by high school students and others for more in-depth research. Public libraries also provide access to the Internet and the wealth of resources found on the World Wide Web.

Academic libraries are found in 2-year technical schools and community colleges, 4-year liberal arts colleges, and in research and land grant universities. The primary focus of academic libraries is to serve the faculties and students of that institution. These libraries are much more specific and comprehensive in the types of information collected, the breadth of subjects covered, and the depth of coverage within specific subject areas. Many academic libraries allow members of the community in which they reside to have access to their library resources. They may or may not allow non-college library users to borrow materials from their library. High school students and teachers at all levels will find the academic library a tremendous information resource for in-depth research and study. Many academic libraries, especially state-sponsored community colleges, 4-year colleges, and universities, encourage high school classes and youth groups to schedule a tour of their library and provide instruction for the use of their library and electronic data products (such as online bibliographic, numeric, and directory databases, electronic journals, and books).

Special libraries are a unique type of library. The services, collections, products, and information provided by special libraries tend to focus on specific data and information needs in the agencies, organizations, and institutions in which special libraries are found. Museums, science and nature centers, federal and most state agencies (few county or municipal) agencies and departments, professional and trade associations, hospitals, civic organizations, special and public interest groups, and businesses and industries are settings in which you are most likely to find special libraries. Many of these libraries are open to the public, especially students, teachers, officials, managers, and leaders of organizations. Because special libraries may have restricted hours of operation or are not staffed with full-time librarians, it is best to call the librarian and ask for their policies concerning public access to their services and collections. Call the agency's or organization's public affairs or public information office to get this same information.

Once you have found a library, do not hesitate to ask the librarian for a brief tour of the facility to determine exactly where specific sections (periodicals, books, reports, government documents, vertical files, card or online catalog, electronic databases, newspapers, reference section) are located. It may take several visits before you become familiar with the location of all these resources.

You should also ask the librarian for help in finding specific information. This can mean instruction in using the card or online catalog of resources the library actually owns or getting a list of articles and then finding the magazines, journals, or newspapers in which they are published. The librarian will show you how to use the Dewey Decimal System, Library of Congress, Superintendent of Documents (U.S. Government Documents), or other library classification systems (especially in special libraries). Special reference books are available to help you define your topic (dictionaries and glossaries), find preliminary general background information (encyclopedias), summaries (abstracts), factual information (handbooks and yearbooks), or inventories (directories, bibliographies, indexes). Librarians can show you how to use these special resources for your environmental information needs.

Another aspect of guidance your librarian provides is instruction on using the Internet and the World Wide Web *effectively and efficiently*. The Internet brings information to the library (and to schools, homes, and offices) in many formats such as text, video, audio, graphics, and animation. The Internet is an extremely powerful information tool and it is important to learn how to use it wisely and responsibly.

The following organizations and their affiliates can be especially useful in finding information about the environment.

American Library Association
http://www.ala.org

    Task Force on the Environment
    http://www.ala.org/alaorg/rtables/srrt/tfoe/

    Libraries Build Sustainable Communities
    http://www.ala.org/sustainablecommunities

    American Association of School Librarians
    http://www.ala.org/aasl/index.html

    Public Library Association
    http://www.pla.org/

    Science and Technology Section of the
      Association of College and Research Libraries
    http://www.ala.org/acrl/sts/sts.html

    Young Adult Library Services Association
    http://www.ala.org/yalsa/

Environmental Protection Agency National Library Network
http://www.epa.gov/natlibra/

Libraries for the Future
http://www.lff.org/

    Environmentalist's Guide to the Public Library
    http://www.lff.org/services/envgui.html

Library of Congress
http://www/loc/gov

Natural Resources Information Council
http://www.quinneylibrary.usu.edu/NRIC/Index.htm

National Agricultural Library
http://www.anl.usda.gov

National Biological Information Infrastructure
http://www.nbii.gov/

National Library for the Environment
http://www.cnie.org/nle

National Library of Medicine
http://www.nlm.nih.gov

Special Libraries Association
http://www.sla.org

    Environment and Resource Management
      Division
    http://www.wco.com/~rteeter/ermd/ermd.html

World Data Centers
http://www.ngdc.noaa.gov/wdc/

    World Data Center-A (United States)
    http://www.ngdc.noaa.gov/wdc/wdca/wdca.html

Reference Works (1990–2000)

Primary literature on the environment includes journal articles, technical reports, government documents, conference papers, data compilations (both print and electronic), and dissertations and theses. These sources of information represent the end results of the research process and make up the body of technical literature about the environment. It is through these primary sources of literature that our knowledge and understanding of the environment is enhanced and advanced. The tasks of identifying, digesting, and selecting pertinent data and information from the large volume of primary literature and putting them into formats and styles that are understood by nontechnical or nonscientific readers is daunting.

The development of a secondary base of environmental literature is needed to facilitate this digestion and understanding of environmental knowledge. This can be achieved by three distinct processes: compacting information, ideas, and theories in encyclopedias, reviews, atlases, and digests; summarizing bibliographic data in bibliographies, indexes, abstracts, and current awareness services; and repackaging data and information in dictionaries, directories, handbooks, yearbooks, and almanacs.

Reference books are designed to be authoritative sources of data, facts, and bits of information. They are rarely written to be read from cover to cover. Learning to make effective use of reference works is an important function for basic library research for laboratory analyses and processes, classroom instruction, field study, paper and report preparation, and other scholarly functions. School and public librarians can be of great assistance in helping to locate reference works to help define, refine, and understand environmental problems and issues.

With the reference process in mind, the following list presents a compilation of atlases, catalogs, dictionaries, directories, encyclopedias, fact books, guides to the literature, handbooks, and yearbooks that focus on the general aspects of the environment, such as ecology, environmental science, environmental studies, global environmental change, and natural resources. These works have been published in the decade of the 1990s with most titles having been written since 1995. Those works followed by an asterisk are noted for students conducting advanced studies or special projects.

Publications

Atlas of Global Change *
Lothar Beckel
Macmillan Library Reference
Simon & Schuster
This 1998 book provides a vivid description of global environmental change. The seventy thematic articles describe various phenomena associated with our changing environment. Desertification, El Niño, climate change, land use, and distribution of resources are among the topics covered in this richly illustrated book.

Beacham's Guide to Environmental Issues and Sources
Beacham Publishing, Inc.
One of the most comprehensive bibliographic assessments of the issues and concerns of the environment for the 1990s and beyond. Published in five volumes with more than 3,200 pages reporting on books, journals, proceedings, and bibliographies in thirty-nine major categories.

The Changing Earth
Facts on File, Inc.
Lavishly illustrated text provides a basic understanding of the key concepts of geochemical cycles and the dynamic nature of our physical environment. This multivolume reference set can be purchased as a whole or by individual titles: *Craters, Caverns, and Canyons: Delving beneath the Earth's Surface; Glacial Geology: How Ice Affects the Land; A History of Life on Earth: Understanding Our Planet's Past; The Human Volcano: Population Growth as Geologic Force; An Introduction to Fossils and Minerals: Clues to the Earth's Past; Marine Geology: Undersea Landforms and Life Forms; Quakes, Eruptions, and Other Geologic Cataclysms; Rock Formations and Unusual Geologic Structures: Exploring the Earth's Surface.*

Conservation Directory
National Wildlife Federation
A comprehensive directory of national, state, and local organizations and agencies; college and university programs; publications; international environmental and natural resources agencies; databases, sources of audiovisual materials, national parks, and so forth. One of the most comprehensive enhanced directories for issues related to the environment and natural resources.
http://www.nwf.org/printandfilm/publications/consdir/index.html

*Conservation and Environmentalism: An Encyclopedia**
Robert Paehlke, editor
Garland Publishing, Inc.
This single-volume encyclopedia provides an international perspective on current concerns about the environment and natural resources. Articles encompass a full spectrum of issues and provide timely discussions of scientific background and policy aspects of many critical issues. Written primarily for the nontechnical audience, it covers broad topics related to environmental quality, resource allocation, and includes contemporary ideas on issues such as toxic chemical hazards, global change, and biological diversity.

Dangerous Weather
Facts on File, Inc.
This is a six-volume reference set providing a fundamental background to studies in climatology, meteorology, and the weather as a driving environmental factor. This set is age-appropriate for middle through high school youth. Entries provide scientific explanations to terms, concepts, and issues. Glossaries, illustrations, and bibliographies are provided. The work can be ordered as a complete six-volume set, or individual titles can be ordered: *Blizzards, A Chronology of Weather, Droughts, Floods, Hurricanes,* and *Tornadoes.*

Dictionary of Ecology, Evolution, and Systematics, 2nd ed.
Roger Lincoln, Geoff Boxshall, and Paul Clark
Cambridge University Press
This extensively revised edition provides more than 11,000 dictionary entries related to biodiversity studies. The topical coverage for this

work transcends the disciplines of botany, zoology, paleontology, natural history, ecology, and environmental studies. Expanded coverage also includes terminology related to global change, soil dynamics, paleoclimatology, ecotoxicology, molecular biology, and genetics.

Dictionary of Environmental Science and Technology
Andrew Porteous
Open University Press
This dictionary provides students (high school and above), community leaders, environmental activists, and other general readers with a working knowledge of scientific and technical concepts and terminology.

Earth Works
Jim Dwyer
Neal-Schuman Publishers, Inc.
This 1997 bibliography guide is subtitled *Recommended Fiction and Non-fiction about Nature and the Environment for Adults and Young Adults.* The book provides descriptions of more than 2,600 popular trade and scholarly books and fictional works in forty-four categories under several major subject areas. An exceptional resource reflecting a high-level effort of the author to provide informative descriptions of numerous works.

Earthright
H. Patricia Hynes
Prima Publishing and Communications
An encyclopedic directory of resources for the environmental activist, teacher, community resource personnel, and others interested in taking action. Nontechnical narratives are complemented with inventories of resources. Broad topical areas addressed include pesticides, solid wastes, drinking water, the ozone layer, and global warming.

Encyclopedia of Environmental Studies
Facts on File, Inc.
A comprehensive source book of more than three thousand environmental entries related to ecology, environmental geography, natural history, environmental law and policy, and environmental management.

Endangered Wildlife and Plants of the World
Marshall Cavendish
Lavishly illustrated with hundreds of maps and four-color photographs. This thirteen-volume set published in 2001 provides details on nearly 1,400 endangered species. Entries are unique in illustrating current endangerment status as reported by both the U.S. Fish and Wildlife Service under the Endangered Species Act as well as the

IUCN—The World Conservation Union. Look for entries on aloes, salmon, clams, tigers, and pine trees, as well as the mysterious coelacanth. Volume 13 provides extensive resources for further study.

Energy Education Resources: Kindergarten through 12th Grade
Paula Altman
National Energy Information Center
This annual report provides a list of generally free or low-cost energy related educational materials available for students and educators. It is updated annually and covers a variety of environmental topics related to energy issues. A PDF file of the full-text document can be obtained at: http://www.eia.doe.gov/bookshelf/other.html

Environmental Disasters
Facts on File, Inc.
A companion to *Man-Made Disasters,* this reference work is devoted to catastrophic episodes that have affected the quality of the environment and relate to the conservation of natural resources. For each disaster the activities and conditions leading up to the event are detailed. The catastrophic or cataclysmic episode is described in terms of short- and long-term effects and impacts. Many of the scenarios read like the unfolding of a suspenseful mystery. Disasters include chemical contamination, deforestation, floods, nuclear events, oil spills, and war crimes against the environment.

Environmental Encyclopedia*
Gale Research, Inc.
This ready-reference source presents a comprehensive, multidisciplinary approach to the study of the environment. The 1997, second edition consists of over 1,300 signed articles and term definitions, providing in-depth, worldwide coverage. This title is part of the Gale Environmental Library, whose titles include: *Gale Environmental Almanac, Gale Environmental Source Book, Encyclopedia of Environmental Information Sources, Statistical Record of the Environment.*

Environmental Grantmaking Foundations
Corinne R. Szymko
Resources for Global Sustainability
The single most comprehensive reference source for identifying grants from public and private independent, community, agency, and foundation sources. Published since 1990, this directory has become a major environmental reference work. This work is provided with extensive indexing to

ease searching for grants in broad topical areas, specific subject categories, geographic considerations (of both the granting body and grant-seeking organization), limitations and uses, and more. Nearly eight hundred individual comprehensive profiles of grantmaking resources are provided. The directory is provided as a CD-ROM product also.

Environmental Guide to the Internet, 4th ed.
Government Institutes, Inc.
The fourth edition of this classic was published in 1997 and provides descriptions of more than five hundred World Wide Web sites and other electronic resources, list serves, electronic journals and newsletters, and other Internet resources.

Environmental Justice: A Reference Handbook
David E. Newton
Contemporary World Issues Series
ABC-CLIO, Inc.
This reference source provides a cogent and comprehensive overview of the issues that have stimulated the current environmental justice movement in the United States. In addition to background information on the development and growth of this environmental concern are chapters devoted to biographical sketches of key individuals in the movement; text of critical laws, government documents, bills, court decisions, and executive orders; inventory of organizations in the field; and bibliographies of relevant books and nonprint resources.

Environmental Key Contacts and Information Sources
Charlene Ikonomou and Diane Pacchione, eds.
Published in 1999 this comprehensive directory provides point-of-contact information for federal government agencies with a detailed section devoted exclusively to the Environmental Protection Agency; state government agencies; nongovernmental organizations: professional, scientific, and trade organizations; publication contacts; books, magazines, journals, and newsletters; and research centers.

Environmental Quality
President's Council on Environmental Quality
The CEQ Annual Report provides an in-depth analysis of trends in environmental quality in the United States and highlights specific aspects of the environment that are improving and the challenges that remain. This yearbook is an essential compendium of data and information related to existing environmental conditions, historical trends, and national policies that impact

the environment. It also provides an update on changes in the National Environmental Policy Act, which in 1970 created CEQ and the environmental impact process. It is widely read by researchers, elected officials, business managers, educators, journalists, students, and those concerned about the quality of the environment. Extensive appendixes of environmental trends data. Also available online as a full-text report.

The Environmentalist's Bookshelf: Guide to the Best Books
Robert Merideth and G.K. Hall
Macmillan Publishing
A compendium of selected environmental books. It is arranged in an interesting fashion to direct the reader to the Top 40 Books. A total of five hundred works are designated as Core Books, Strongly Recommended Books, and Other Recommended Books. The books listed were placed in their respective categories based on survey responses distributed to environmental leaders in various academic, NGO, government, and other positions.

Facts on File Encyclopedia of Garbage
Facts on File, Inc.
A detailed description of the municipal solid waste stream more commonly referred to as garbage. Provides insights on its generation, removal, disposal, and impacts related to environmental health and public policy. Industrial waste and chemical effluents and airborne releases are also included.

Facts on File Dictionary of Environmental Science
L. Harold Stevenson and Bruce Wyman
Facts on File, Inc.
A comprehensive, multidisciplinary dictionary for neophytes and seasoned veterans seeking definitions for critical terms and concepts related to environmental quality, occupational health, and wildlife conservation.

Facts on File Environment Atlas
Facts on File, Inc.
Geography, ecology, natural history, and environmental studies are interwoven into the landscapes that comprise our environment: wildscapes, farmscapes, and townscapes. The environment is broadly defined from pristine wilderness areas to the heart of urban centers. Provides general background information and descriptions of critical concepts, issues, and topics related to the environment and natural resources. Liberally illustrated text provides middle school and junior high students with the basic concepts of environmental science.

Facts on File Wildlife Atlas
Facts on File, Inc.
The complexities of ecology unfold with concise descriptions of natural habitats, biogeochemical cycles, and strategies of conservation related to different types of plants and animals found in natural environments.

Global Ecology Handbook
Walter H. Corson, ed.
Global Tomorrow Coalition
A comprehensive directory of resources and organizations that ushered in the environmental movement of the 1990s. Arranged by 16 topical areas related to energy, population, development, environmental quality, and natural resources.

*Global Environmental Outlook**
United Nations Environment Programme
Oxford University Press
This compendium is the first report of the Global Environment Outlook Project and reviews major environmental issues from regional and international perspectives. This report also provides a summary and evaluations of policy responses addressing a variety of environmental concerns.

*Great Events from History: Ecology and Environment Series**
Frank N. Magill, ed.
Salem Press
A five-volume treatise on the history of ecology and the environment from 1902 (the Reclamation Act) to 1995 (date of improvement in the official endangerment status of the bald eagle). The time span for each volume is as follows: Vol. 1:1902-1944, Vol. 2:1945-1966, Vol. 3:1966-1973, Vol. 4:1973-1985, and Vol. 5:1985-1994. Each event listed in this chronological compendium is provided a summary essay, principal personages mentioned, a brief bibliography, and a list of cross-referenced essays

Jobs You Can Live With: Working at the Crossroads of Science, Technology, and Society
Susan M. Higman, ed.
Student Pugwash USA
A directory of organizations working for a better world. The mission of Student Pugwash USA is to promote socially responsible applications of science and technology in the twenty-first century. This directory is for students and young professionals who are in the midst of their academic studies or are just beginning their careers and want to start their life's work by creating a more just, secure, and sustainable world. A tremendous resource for identifying internship opportunities.

Man-made Catastrophes
Facts on File, Inc.
A comprehensive encyclopedia of disasters of human origin from around the world. Covers events from the earliest recorded times to the present

The Official World Wildlife Fund Guide to Endangered Species of North America
Beacham's Guide to International Endangered Species
Beacham Publishing, Inc.
September 1, 1914, was the date on which Martha, the last surviving passenger pigeon, died alone in a cage in the Cincinnati Zoo. Flocks of these birds were so plentiful in the mid-1800s that the species may have been the most plentiful in North America. This reference series attempts to prevent stories like Martha's from happening. This dictionary provides a comprehensive inventory of plant and animal species facing the threat of extinction in North America. The lists are derived from the U.S. Fish and Wildlife Service's publication, "Endangered and Threatened Wildlife and Plants" and its supplements, information from the Federal Register, and the "Endangered Species Technical Bulletin," and other official government documents and reports from the international community of endangered species experts in public and private sectors.

Our Changing Planet
GCRIO User Services
Global Change Research Information Office
A yearbook providing an annual update of the achievements and projected goals of the U.S. Global Change Research Program. In addition to project accomplishments and program summaries are descriptions of funding support for the Global Change Research Program. An essential resource for researchers, policy makers, educators, and students.

Planet Management: The Illustrated Encyclopedia of World Geography
Michael Williams
Oxford University Press
This reference work looks at environmental issues from a regional perspective and directs readers to the technologies, policies, and strategies needed to address and remedy problems. Richly illustrated with maps, this work is an environmental atlas with crisp textual descriptions and narratives providing an encyclopedic touch to the work. Topics include issues related to population, environmental quality, resource management and realignment of technologies, policies, and social paradigms. Lavishly illustrated with more than 270 color photographs.

Reading about the Environment
Libraries Unlimited
This book is designed to help librarians and other readers to locate information currently available about the environment. This annotated bibliography describes nearly eight hundred books, popular magazine articles, and other materials found in many public libraries.

State of the World
Worldwatch Institute
An annual desktop reference book used by government officials, researchers, corporate managers, journalists, educators, students, and others concerned about a wide range of environmental issues. This yearbook is one of the most widely used resources that bridges scientific and policy aspects on key environmental issues. Important environmental data are included. Available also as a CD-ROM.

There Goes the Neighborhood: Cartoons on the Environment
Sidney Harris
The University of Georgia Press
Sidney Harris's cartoons provide a humorous and often poignant contribution to the discussion of science and technology. His works are regularly featured in magazines and journals such as *The New Yorker* and *American Scientist*. For many years Harris's works were featured in the Institute for Scientific Information's weekly publication, *Current Contents*. This 1996 collection is devoted to the topics of ecology, natural resources, waste, and the environment.

Vital Signs
Worldwatch Institute
The subtitle of this book is, "Environmental Trends That Are Shaping Our Future." This annual summary describes the trends related to the major environmental issues facing the world: Food, Agriculture, Energy, the Atmosphere, Economics, Transportation, Social Issues, and Military Trends. Special features on specific key issues within these categories are also provided.

Wetlands
Facts on File, Inc.
A thorough, continent-by-continent inventory of wetland areas with color photographs, maps, and line drawings.

World Resources
World Resources Institute
A biennial (every-other year) comprehensive review of the critical issues and challenges facing the world's environmental leaders. Provides surveys of current conditions and historical trends in major areas of concern: Economic Indicators, Population, Forests and Land, Food and Agriculture, Biodiversity, Energy and Materials, Water and Fisheries, and Atmosphere and Climate. Extensive data tables are provided. Data also available in diskette formats.

World's Who is Who and Does What in Environment and Conservation
Nicholas Polunin, ed.
St. Martin's Press
A biographical reference work, published in 1997, providing descriptions of more than three thousand notable people working in the fields related to environmental science, ecology, natural resources conservation, and wildlife biology. Contact information (mailing addresses, phone, fax, etc.) is provided where made available.

Providers of Reference Information

The following are important providers of reference information in a variety of formats:

Association for Biodiversity Information
2445 North Fairfax Drive, Suite 400
Arlington, VA 22203-1606
703/841-7195 (tel)
703/525-8024 (fax)
mschaefer@tnc.org
http://www.abi.org/abistaff.htm

Global Change Research Information Office
P.O. Box 1000 — 61 Route 9W
Palisades, NY 10964
914/365-8930 (tel)
914/365-8922 (fax)
help@gcrio.org
http://www.gcrio.org (full text available)

Government Institutes, Inc.
4 Research Place
Rockville, MD 20850
301/921-2355
giinfo@govinst.com
http:www.govisnt.com

National Energy Information Center
U.S. Department of Energy
1000 Independence Avenue, S.W.
Washington, DC 20585
202/586-1175

National Wildlife Federation
1400 16th Street, N.W.
Washington, DC 20036-2266
202/797-6800

President's Council on Environmental Quality
722 Jackson Place, N.W.
Old Executive Office Building, Room 360
Washington DC, 20502
202/456-6224
http://www.whitehouse.gov/WH/EOP/CEQ/html
/CEQ.html

Student Pugwash USA
815 15th Street, N.W.
Washington, DC 20005
202/393-6555
202/393-6550 (fax)
spusa@spusa.org
www.spusa.org/pugwash/

World Resources Institute
10 G Street, NE (Suite 800)
Washington, DC 20002
202/729-7600
http://www.wri.org/

Worldwatch Institute
1776 Massachusetts Avenue, N.W.
Washington, DC 20036-1904
202/452-1999

Internet Resources

In the past ten years the evolution and growth of the Internet and the World Wide Web has revolutionized the ways to identify, obtain, use, share, and store information. What started as a cold war technology in 1969 (the Defense Advanced Research Projects Agency Network, or DARPANet, was the prototype of the Internet) has emerged today as a mainstay for getting information of any type from one place to another. Communicating has taken on new dimensions. The following Internet and World Wide Web resources will assist environmental educators and their students in all formal K-12 schools as wells as other institutions such as nature centers, museums, youth camps, and related organizations.

The environment and natural resources are areas that have taken full advantage of the power of the Internet and the World Wide Web. Toni Murphy and Carol Briggs-Erickson's Environmental Guide to the Internet is listed above in the section on environmental reference books. Not only does this book provide more than 350 pages of environmental and biogeochemical sources of information on the Internet, it provides a brief history of the creation of the Internet and the World Wide Web. There is an outstanding subject index provided for this book.

Tim Hopkins has written a good reference book for teachers, *1001 Best Websites for Educators,* with a CD-ROM for linking directly to Web sites listed. He has devoted a chapter to Geography/Environment. There are a number of metasites or online gateways that attempt to organize the vast array of data and information about the environment. The following is a brief list of some of the more useful ones.

The Argus Clearinghouse, http://www.clearinghouse.net/, is a central place where value-added topical guides on the environment can be found. Unique aspects and benefits of this site are that subject specialists have created the individual sites listed and staff at the Argus Clearinghouse proved evaluations and rankings based on quality for the Internet-based information resources listed. The Environment category includes five sections: ecology (with 12 subheadings), environmental activism (3), environmental law (2), sustainable development (6), and waste management (2). In the Science and Mathematics section you will find an inventory of Web sites related to the earth sciences (23). The Recreation section has a category of nature activities (8).

The WWW Virtual Library—Environment, http://earthsystems.org/Environment.shtml, is an online inventory of Web sites on environmental topics maintained by earthsystems.org, a nonprofit environmental education organization. Within this site are resources listed for 60 different categories from Academic Programs to Weather.

Other related Virtual Library sites are:

Biodiversity & Ecology
http://conbio.rice.edu/vl/

Earth Sciences
http://www.geo.ucalgary.ca/VL-EarthSciences.html

Education for Sustainable Development
http://esdtoolkit.org/

Energy
http://gem.crest.org/

Environmental Law
http://www.law.indiana.edu/v-lib/

Forestry
http://www.metla.fi/info/vlib/Forestry/

Landscape Architecture
http://www.clr.toronto.edu/VIRTUALLIB/larch.html

Meteorology
http://dao.gsfc.nasa.gov/DAO_people/towens/VLM/

Paleoclimatology and Paleo-oceanography
http://www.datasync.com/~farrar/www_vl
_paleoclim.html

Science Fairs
http://physics.usc.edu/~gould/ScienceFairs/

Sustainable Development
http://www.ulb.ac.be/ceese/meta/sustvl.html

Urban Environmental Management
http://www.gdrc.org/uem/

Skewlsites lists quality educational Web sites based on criteria such as content, appearance, and ease of navigation. Maintained and directed by teachers, this site also offers a monthly newsletter with new additions to the page. Among its multidisciplinary listings are Environmental Sciences, http://www.skewlsites.com/envirsci.htm, and Earth and Space, http://www.skewlsites.com/ethsp.htm.

Individual college and universities develop and maintain Internet gateways as subject specific inventories of electronic and digital resources. Many college and university OPACs (online public access catalogs) are now indexing major Web sites. Many OPACs (e.g., the University at Buffalo's

BISON http://ublib.buffalo.edu/) are searchable by anyone in the world at no cost. Libraries and departments within colleges and universities also provide free access to their subject-specific compilations of environmental information, such as the Environment site maintained by the Science and Engineering Library at the University at Buffalo, http://ublib.buffalo.edu/libraries /units/sel/environ/environ.html.

The Yahoo Environment site, http://dir.yahoo.com /society_and_culture/environment_and_nature/, is another very fertile source of Internet resources about the environment and nature.

The URLs (Uniform Resource Locator) in this section have been tested and are the most recently accessible Web addresses available at the time of writing. Internet and Web addresses change frequently. If a site is no longer operating or its address has changed, go to an Internet search engine and enter the name of the site you need and see what happens. The University at Buffalo Undergraduate Library maintains a good, partial list of Web search engines and a tutorial for searching the Internet for information, http://ublib.buffalo.edu/libraries /search/searchint.html.

Environmental Education Information Providers

The Environmental Education and Training Partnership Resource Library at Ohio State University Extension (700 Ackerman Road, Suite 235, Columbus, OH, 43202-1578; tel: 614/292-6926; e-mail: heimlich@osu.edu) has published a Directory of Environmental Education Information Providers. Each organization listed is provided a summary statement of resources, services, funding, publications, and other information related to their abilities to provide scientifically accurate information on any aspect of the environment.

Center for Environmental Education
http://www.cee-ane.org/index.html

Center for Great Lakes Environmental Education
http://www.greatlakesed.org

Eisenhower National Clearinghouse
http://www.enc.org

The Environmental Media Corporation
http://www.envmedia.com

ERIC (Education Resources Information Clearinghouse) Clearinghouse for Science,

Mathematics, and Environmental Education
www://www.ericse.org

GLOBE
http://www.globe.gov

The Institute for Global Environmental Strategies
www.strategies.org

The North American Association for Environmental Education (NAAEE)
http://www.naaee.org or
http://eelink.umich.edu/naaee.html

National Council for Science and the Environment (formerly the National Institute for the Environment)
http://www.cnie.org

Second Nature
http://www.2nature.org

World Resources Institute
http://www.wri.org

Additional Links for Environmental Education

Additional resources and expertise on environmental education can be obtained from the following. These sources should be consulted for general overviews of resources and services related to environmental education. Many of them have extensive inventories of resources, including links to other Web sites, searchable databases, publication lists, newsletters, discussion lists, and related resources.

Earth Sense
http://www.atmos.uah.edu:80/essl/earthsense/
A hydrology and remote sensing education project for K-12 earth science teachers.

Educational Resources for Cartography, Geography, and Related Disciplines
http://www-nmd.usgs.gov/www/html/1educate.html
From the USGS National Mapping Information Program, this site provides teachers with resources for introducing geographic concepts that are critical for learning about global climate change.

EE-Link Environmental Information on the Internet
http://eelink.net/sitemap.html
EE-Link is a project of the Global Rivers Environmental Education Network (GREEN) and a participant in the Environmental Education and Training Partnership of the North American Association for Environmental Education. Contact them at the GREEN office, 206 S. 5th Ave., Suite 150, Ann Arbor, MI, 48104; tel: 734/761-8142; fax: 734/761-4951.

Energy Education Resources: Kindergarten through 12th Grade
http://www.eia.doe.gov/bookshelf/other.html
This report is compiled by the National Energy Information Center, a program of the U.S. Department of Energy's Energy Information Administration. Energy Education provides a list of generally free or low-cost educational materials available for students and educators. It is updated annually and covers a variety of environmental topics related to energy issues, including those related to global climate change.

K-12 Geoscience Education Resources
http://www.geo.arizona.edu/K-12/outreach.html
General resources for geoscience education.

National Wildlife Federation Environmental Education
http://www.nwf.org/nwf/education/
NWF is a leader in environmental education, reaching out to the teachers, parents, and communities where children and adults live, work, and play.

Oak Ridge Institute for Science and Education (ORISE)
http://www.orau.gov/orise/educ.htm
ORISE provides educational opportunities for high school teachers and students, undergraduate and graduate students, post-graduates, and faculty.

Science
http://www.middleweb.com/CurrScience.html
Articles, e-mail, and Web links about science, primarily for middle, junior-high, and high school students.

World Resources Institute (WRI) Environmental Education
http://www.wri.org/
WRI presents a wealth of resources in many formats from full-text online access to videos, slides, and guides that include lesson plans, overhead transparency masters, and student exercises.

Environmental Education (EPA)
http://www.epa.gov/ocepa111/NNEMS/index.html
This is the EPA's official environmental Web site, created by the Office of Environmental Education. This site is primarily for educators with links for resources in training, funding, partnerships, student opportunities, and resources.

The National Environmental Education and Training Foundation
http://www.neetf.org/
The mission of the NEETF is to help America meet critical national organizations on voluntary initiatives that supplement regulation with innovative compliance and cooperation.

Links on Global Environmental Change

Some of the most active environmental education resource development is taking place with resources related to global environmental change, including climate change and global warming. The following is a list of education efforts on this timely and important environmental topic.

Center for Earth Observing and Space Research
http://www.ceosr.gmu.edu/
Provides data and information resources from NASA's Earth Observing System (EOS) beyond the traditional audiences of researchers and brings these science-based resources to teachers and educators at all levels from middle school through college. For additional NASA education resources, see also NASA's Online Educational Resources site at http://quest.arc.nasa.gov/OER/

Climate Solutions: Growth with Less Energy
http://www.panda.org/climate/solutions/
This site looks at achieving the reduction of greenhouse gases to stop the warming of the global climate. It provides links to discuss the science, business, and policies related to global climate change.

College and University Global Change Courses
http://gcrio.gcrio.org/edu/highered.html
The course descriptions, syllabi, and course home pages listed here demonstrate the range of global environmental change studies available at colleges and universities. More than one hundred individual classes are listed. Many can be adapted to or modified for inclusion in high school science and Advanced Placement classes.

Destination Earth
http://www.earth.nasa.gov/
The official Web site of NASA's Earth Science Enterprise, a comprehensive education initiative to increase students' knowledge about the sciences related to the earth. Earth Science Enterprise (NASA) http://www.hq.nasa.gov/office/mtpe/edreports.html, reviews various NASA education products related to earth science or earth-observation technology. Includes links to full-text copies of ESE Education Reports (overviews) beginning with May 1995.

Earth Science Resources: Geology, Oceanography, Astronomy, and Ecology
http://jrscience.wcp.muohio.edu/EarthSci.html
Provides resource links for educators on many topics related to global environmental change.

Education for Sustainable Development Tool Kit
http://www.esdtoolkit.org
The purpose of the ESDTool Kit is to help schools and communities develop a process to create locally relevant and culturally appropriate education. The tool kit is based on the idea that communities and educational systems within communities need to dovetail their sustainability efforts. Local educational systems can reorient existing curriculums to reinforce local sustainability goals.

Electronic Mailing Lists for Global and Environmental Change Education
http://gcrio.org/edu/edu.mailing.html
Discussion lists compiled by Global Change Research Information
(http://gcrio.org/edu/edu.mailing.html).

GCDIS: Educational Resources
http://www.globalchange.gov/cgi-bin/gcdis-query-ssi2?education
An inventory of more than thirty education sites for global change studies. Accessed from the Global Change Data and Information System Home Page, http://www.globalchange.gov/, by clicking on the Educational Resources box. Includes several very good resources for geographic and cartographic education resources.

GCRIO Unit Lesson Plans
http://gcrio.gcrio.org/lp/biodiv/biodiversity.html
Provided by the U.S. Global Change Research Information Office. Unit One: International Environmental Treaties for Conserving Biological Diversity (high school and college level). This unit introduces biological diversity and the importance of international environmental treaties in sustainable living. Students can experience the decision-making process involved in the development and implementation of environmental conventions; develop ecological literacy; and see the need for science in consideration of global environmental issues. Teachers can access an online biodiversity module that provides an overview, glossary, facts, activities, additional resources, hands-on exercises, and minds-on discussion topics.

The Global Change Game
wysiwyg://194/http://www.gcg.mb.ca/
Played on a world map the size of a basketball court. A great learning activity that stimulates creative problem-solving and critical thinking strategies.

Global Change and Environmental Education
Resources
http://gcrio.gcrio.org/edu/educ.html
A multidisciplinary and international collection of
resources representing a wide range of resources in
a variety of formats for educators and students at
all levels (K-12 and higher education), librarians,
citizens, and community groups.

Global Change and Stability
http://www.vims.edu/adv/gc/gcei.html#top
Produced by the Virginia Sea Grant Global Change
Education Program, this site includes An Educators'
Page with links to teaching resources, lesson plans,
and activities. Student global change papers are
presented, and a Global Change and Sustainability
Index with reading lists, overviews, and fact sheets is
provided. Sea Grant regional global change
education programs are provided for the Mid-
Atlantic states (Delaware, Kentucky, Maryland, New
Jersey, North Carolina, Virginia, West Virginia, and
Tennessee) and the states bordering the Great Lakes
(Illinois, Indiana, Michigan, Minnesota, New York,
Ohio, Pennsylvania, and Wisconsin); see
http://www.greatlakes.net/education/educate.html

Global Climate: A Resource for Education
http://winds.jpl.nasa.gov/education/eduindex.html
The Scatterometer Program Office is committed to
meeting the increasing requests for educational
resources about NASA's Space and Earth Science
activities.

Global Learning and Observations to Benefit the
Environment (GLOBE)
http://www.globe.gov/
The National Oceanographic and Atmospheric
Administration (NOAA) supports the GLOBE
Program, a worldwide network of K-12 students,
scientists, and teachers working together to learn
more about the global environment.

Global Warming: Focus on the Future
http://www.enviroweb.org/edf/
This is a mini-tour of the full global-warming
exhibit on permanent display at the Biosphere 2
Center outside Tucson, Arizona. A rather extensive
inventory of resources.

Global Warming and Climate Change
http://gcrio.org/gwcc/toc.html
This brochure was prepared by researchers at
Carnegie Mellon University to explain the issue of
global warming and climate change. Provides
background information for teachers at all levels.

Institute for Global Change Research and Education
http://space.hsv.usra.edu/~igcre/
IGCRE is located at the Global Hydrology and Climate
Center, an institution jointly operated by the
University of Alabama at Huntsville and Universities
Space Research Association. IGCRE collaborates with
the NASA/Marshall Space Flight Center's Earth System
Science Division to focus on advancing the
understanding of the role of water and energy in the
dynamics of global change. In addition, the institute is
integrating research with the educational needs of
earth system and global change science, including
those at the pre-college level, to convey the importance
and exciting challenges of global change science.

International Research Institute for Climate Prediction
http://iri.ucsd.edu/
A cooperative effort of the National
Oceanographic and Atmospheric Administration's
(NOAA) Office of Global Programs, Columbia
University's Lamont-Doherty Earth Observatory,
and the Scripps Institution of Oceanography has
established this International Research Institute
(IRI) to provide experimental climate-forecast
guidance on seasonal-to-interannual time scales
for use by affected communities around the world.

IT'S ELEMENTARY!
http://gcrio.gcrio.org/edu/elementary/itselem.html
A teacher designed this Web site specifically for
teachers to access, view, and print lessons during
one planning period using their school's Internet
connection. For K-8 teachers.

JASON Project
http://www.jasonproject.org/
A comprehensive environmental change project based
on scientific inquiry and the thrill of exploration.
Resources and programs for teachers and students.

Sierra Club Global Warming Campaign
http://www.sierraclub.org/global-warming/
This site provides news and video links for teachers
and students.

Supportive Scientific Visualization Environments
for Education (SsciVEE) Project
http://www.worldwatcher.nwu.edu/sciviz.html
Scientific visualization has had a tremendous
impact on the practice of science during the past
decade by capitalizing on the power of the human
visual-perception system to identify patterns in
complex data. The SsciVEE project is exploring the
potential of this technology to improve science
education in similar ways.

A Weather Page for Kids, Teachers, and Parents
http://nimbo.wrh.noaa.gov/Portland/edukids.html
Listings of resources and activities for children to use to explore weather and increase their interest and understanding of science with projects for the classroom and home.

Links for Other Science-Related Resources

The following educational resource sites provide a wide variety of extremely useful links to further resources for science education.

Hot Links!
http://128.32.190.250/HotLinks.html

Monterey Bay National Marine Sanctuary Education Program
http://bonita.mbnms.nos.noaa.gov/Educate/

Online Education: Geostationary Satellites
http://goeshp.wwb.noaa.gov/links.html

Online Educational Resources (NASA)
http://quest.arc.nasa.gov/OER/

Rainforest and Reef
http://www.rainforestandreef.org/

Science Fairs and Forums
http://mel.lib.mi.us/science/fairs.html

The Weather Learning Center
http://www.intellicast.com/wxshops/wxclass.htm

WINDandSEA: The Oceanic and Atmospheric Sciences Internet Locator (NOAA)
http://www.lib.noaa.gov/docs/windandsea.html

World Resources Institute Environmental Education Project
http://www.wri.org/enved/

Professional Science Associations

The following professional associations have resources, newsletters, publications, workshops, Web site inventories, and other resources that can be of benefit to environmental educators and their students.

American Association for the Advancement of Science: Project 2061
http://project2061.aaas.org/

American Association of Physics Teachers
http://www.aapt.org/

American Chemical Society: Science Education
http://www.acs.org/govt/issues/2st34.htm and http://www.acs.org/education/currmats/n01.html

National Association of Biology Teachers
http://www.nabt.org/

National Association for Research in Science Teaching
http://science.coe.uwf.edu/narst/narst.html

National Association for Science, Technology, and Society
http://www.tcnj.edu/~nasts/nasts1.htm

National Institute for Science Education
http://www.wcer.wisc.edu/nise/

National Science Education Leadership Association
http://science.coe.uwf.edu/NSELA/NSELA.html

National Science Teachers Association
http://www.nsta.org/

New York State Outdoor Education Association
http://www.nysoea.org/

North American Association for Environmental Education
http://www.naaee.org/

Pittsburgh Regional Center for Science Teachers
http://www.pitt.edu/~sosmos/

Triangle Coalition
http://www.triangle-coalition.org/

Databases and Audiovisual Media

Audiovisual materials have a long and popular place in classroom settings. However, the availability of such resources in videotape, audiotape, CD-ROM, and DVD formats is a recent phenomenon that has taken audiovisual presentations into summer camps, family rooms, living rooms, and even commercial movie theaters. This section is not to be viewed as an exhaustive inventory of resources, but rather as a pointer for students, teachers, parents, and mentors to selected databases and sources of audiovisual media.

Reference resources for locating audiovisual materials include the National Wildlife Federation's annual *Conservation Directory* (Vienna, VA: National Wildlife Federation), Bowker's *Complete Video Directory* (New Providence, NJ: R.R. Bowker), and *Video Source Book* (Detroit, MI: Gale Research). SB&F (Science Books & Films), published nine times per year by the American Association for the Advancement of Science, http://www.aaas.org, provides librarians, media specialists, curriculum supervisors, science teachers, and educators with critical reviews of the scientific accuracy of print, audiovisual, and electronic resources, including those related to ecology, the environment, and natural resources.

The book and gift stores of nature centers, science museums, botanical gardens, and museums of natural history listed in this section and elsewhere are very good sources of audiovisual resources and classroom aides. Do not forget to use the library catalog (hard copy card catalog or the electronic online public access catalog) to find audiovisual resources in your local school, public, and other libraries.

Many of the government agencies provided in this section maintain extensive resources for educational purposes, including posters, CD-ROMs, PowerPoint presentations, charts, and in some cases video and audiotape. Take the time to explore the education sites developed by federal, national, state, and provincial agencies for access to audiovisual, nonprint, and nontext resources.

Databases

Check with your school or public library for availability.

Catalog of U.S. Government Publications
U.S. Government Printing Office
732 N. Capital St.
Washington, DC 20402
202/512-1530
http://www.access.gpo.gov/su_docs/locators/cgp/index.html
The catalog is a search and retrieval service providing bibliographic records of U.S. Government information products, including audiovisual resources. Use it to link to federal agency online resources or identify materials distributed to Federal Depository Libraries.

LOCATOR plus
U.S. National Library of Medicine
8600 Rockville Pike
Bethesda, MD 20894
888/FIND-NLM
301/594-5983
http://www.nlm.nih.gov/locatorplus
Contains bibliographic records of audiovisual and nonprint materials cataloged by the National Library of Medicine. Search LocatorPlus; click "Limits" icon; in "Item Type" or "Medium" menu bars select appropriate materials or resources (or other limitations) sought and click "Set Limits" task bar; enter search terms (e.g., "air pollution"), retrieve results. Replaces NLM's AVLINE database. Many titles may be of a more technical nature.

National Information Center for Educational Media (NICEM)
P.O. Box 864
Albuquerque, NM 87198-8640
505/256-1080
mailto:nicem@nicem.com
http://www.nicem.com
Bibliographic data of videos, audiotapes, software, filmstrips, slides, and assorted educational media through a variety of means including online, CD-ROM, audio, graphic, and print. Requires subscription. Check with your public or school library, or if possible a local college or university library for availability.

WorldCat
Online Computer Library Center (OCLC)
6565 Frantz Road
Dublin, OH 43017-3395
614/764-6000
mailto:oclc@oclc.org
http://www.oclc.org
A megalibrary catalog containing more than 45 million records contributed by 20,000 libraries around the world. Database can be searched by media type (visual materials, sound recordings, and mixed media). This database requires a subscription. Check with your public or school library, or if possible a local college or university library for availability or for comparable resources.

Videotapes

A major factor influencing the popularity of electronic and digital formats is the demand by students, teachers, parents, and youth leaders for quality information in formats that not only communicate ideas and teach concepts, but resources that do so in an entertaining manner. Such entertainment factors are even realized in major full-length feature movies such as *Never Cry*

Wolf (1983), *The River* (1984), *Medicine Man* (1992), *Outbreak* (1995), *Kayla: A Cry in the Wilderness* (1997), *A Civil Action* (1998), *Erin Brockovich* (2000), *Wolves* (IMAX, 1999), *Dolphins* (IMAX, 2000), *Whales* (IMAX, 2000), and *Bears* (IMAX, 2001).

Another source of environmental videos are those produced as made-for-TV movies, such as the specials produced by the Discovery Channel, PBS (e.g., Nature, Nova, Scientific American Frontiers), and National Geographic. These videos are often available for sale and in many instances can be rented from local video stores or borrowed from school or public libraries. A list of major contributors to television coverage of the environment and nature includes:

Cable News Network (CNN)
http://www.cnn.com/

Discovery Channel
http://www.discovery.com/

ESPN: Trout Unlimited TV
http://www.tutv.org/

The Outdoor Channel
http://www.outdoorchannel.com/

Public Broadcasting System
http://www.pbs.org/

The Learning Channel
http://tlc.discovery.com/

WGBH-TV: NOVA
http://www.pbs.org/wgbh/nova/

The Weather Channel
http://www.weather.com/

Companies that produce and distribute videos (and more recently DVDs) specifically related to the environment, natural resources, and wilderness themes are listed below. Each provides free catalogs of their offerings (many of which are found in public and school library collections). Following is only a small sample of available distributors and titles.

Bullfrog Films
P.O. 149
Oley, PA 19547
800/543-3764
bullfrog@igc.apc.org
http://www.bullfrogfilms.com/

Gaia: The Living Planet
Once and Future Planet
Our Planet Earth

Carolina Biological Supply Company
2700 York Drive
Burlington, NC 91311
800/334-5551 (U.S.)
336/584-0381
carolina@carolina.com
http://www.carolina.com/
Earth Sciences
   Discovering Our Earth's Atmosphere
   Discovering Our Planet Earth
   El Niño and the Link between Oceans,
      Atmosphere, and Weather
Ecology
   A Walk through the ACEER (Amazon Center
      for Environmental Education and
      Research)
   Useful Plants Trail
   A World Alive
   4,000 Meters under the Sea

edudex.com
P.O. Box 8197
Princeton, NJ 08543-8197
877/301-3100
609/720-1234
http://www.edudex.com/
edudex.com is a multidisciplinary marketplace for educational materials, including videos, software, and audio products. From the home page go to the "Browse Category." In the Environmental Science category you will find the topics Conservation/Endangered Species, Ecology, Ecosystems, Energy/Natural Resources, Environmental Science-General, Global Warming, Pollution, Population, Rainforests, and Waste Management/Recycling.

Environmental Media
P.O. 99
Beaufort, SC 29901-0099
800/ENV-EDUC
General environmental media resources in print, video, and CD-ROM formats.

Films for the Humanities and Science
P.O. 2053
Princeton, NJ 08543-2053
800/257-5126
http://www.films.com
   Earth Story (8 part series)
   Noah's Children (covers climate changes
      throughout history)
   El Niño: Disaster on the Wind

The Fallacy of Global Warming
Ocean Science: Worlds Colored Blue
The Restless Planet
Understanding Energy
The World's Weather

Hawkhill Video
125 East Gilman Street
P.O. 1029
Madison, WI 53701-1029
800/422-4295
www.hawkhill.com
Biosphere
The Changing Earth
Chemical Cycles in the Biosphere
Climate, Weather, and People
Climate and Weather
Earth Scientists at Work
Ecosystems
Energy and Society
Energy for Tomorrow
Global Warming
How Serious is Global Warming?
Spaceship Earth

Media Basics Video
Lighthouse Square
P.O. 449
Guilford, CT 06437-0449

800/458-2505
203/458-2505
Carbon Cycle
Earth Watch
Energy in Our Environment
Fossil Fuels
The Greenhouse Effect
The Hydrologic Cycle: Water in Motion

National Wildlife Federation
800/278-7599
http://www.nwf.org/productions/onvideo.html
Return of the Eagle
Survival of the Yellowstone Wolves
Tiger!

The Video Project
200 Estates Drive
Ben Lomond, CA 95005
800/4-PLANET
408/336-0160
www.videoproject.org
Secrets of Science Series (3 sets, 13 volumes)
Polluting Our Atmosphere
Creating Our Climate
Greenhouse Crisis: The American Response
Ozone: Cancer of the Sky

Slides and Transparencies

Ecology-related slides and transparencies are available from Carolina Biological Supply (see above). From their online catalog scroll down to the "Transparencies, 35mm" category and select ecology. Sets are available for *The Niche and Ecosystem, Coral Reef Community, Food Chains and Pyramid, Pond*

Food Web, Biomes, Ecological Succession, and *Symbiosis.* Many slide-set presentations have gone the way of the dinosaur and are being replaced by online equivalents. Scout the Internet and World Wide Web for PowerPoint presentations or slide, photograph, or image archives.

Posters and Wall Charts

Carolina Biological Supply (see above) is one of the leading providers of educational wall charts and posters in the biological sciences. From their online catalog one goes to the "Charts" bar and finds a variety of biology-related resources, including topics related to: earth science (Clouds and Weather) and environmental science (Air Pollution, Barrier Islands, Caribbean Coral Reef, Chesapeake, Noise Pollution, Ocean Forest, Pollution Set, Rainforest, Rainforests of the World Set). Fisher Scientific, http://www3.fishersci.com/, and other suppliers of classroom equipment, supplies, and resources generally distribute audiovisual resources.

Many students, teachers, and parents can also use artistic and esthetic posters to stimulate interest in the environment and nature. The following sites represent resources of art posters related to the environment and natural resources. Be sure to navigate around these respective Web sites for other nature, wildlife, and environmental-themed posters.

Earth
http://www.allwall.com/asp/display-asp/_/NV—1_33_36_134/1.asp

Earth Day Posters
http://students.resa.net/stoutcomputerclass/clubposters.htm

Earth Image Ultimate Image of the Earth
http://www.earthimage2000.net/gallery.htm

Earth Image Earth Day 2000 Contest Winners
http://www.earthimage2000.net/winners_poster
_all.htm

The Earth Is a House
http://www.allaboutart.com/detail.asp?site=aaa&
title=earth&page=1&pos=8

The Earth Is My Mother
http://www.greenwichworkshop.com/bev_feature2
.html

Earth Week Posters
http://www.fcps.k12.va.us/ChurchillRoadES
/crs9798/earthday/posters/

Fragile Earth
http://www.geology-net.com/bpe00-04.htm

Earth Day 2000 Poster
http://www.dep.state.pa.us/earthdaycentral/poster
.htm

Planet Earth
http://www.concertposter.com/p/pcplpe1042p.htm

Audiotapes

In addition to resources described above, there are a variety of organizations that solely distribute audiotapes related to environmental quality and natural resources. Most notable among these groups in the United States is National Public Radio (NPR), http://www.npr.org. NPR produces a variety of radio programs related to the environment. These programs provide tapes and transcripts of their broadcasts and more recently delayed RealAudio coverage of the actual broadcasts for playback at later times. Among NPR and other public radio programs of a special science or environmental nature are the following.

All Things Considered
http://npr.org/programs/atc/
(search for environmental topics)

Earth News Radio (Environmental News Network)
http://www.enn.com/radio/

Living on Earth with Steve Curwood
http://www.loe.org/

National Press Club
http://www.npr.org/programs/npc/
(search for environmental topics)

Pulse of the Planet
http://www.pulseplanet.com/

Radio Expeditions
http://www.npr.org/programs/RE/
(search for environmental topics)

Sounds Like Science
http://www.npr.org/programs/science/
(search for environmental topics)

Talk of the Nation—Science Friday with Ira Flatow
http://www.npr.org/programs/scifri/
(search for environmental topics)

The Weather Notebook
http://www.mountwashington.org/notebook/
index.html

In Canada, the following public radio and television programs should be consulted for coverage of environmental and natural resources themes. Be sure to use the sites' search features to locate broadcasts, archives, and scripts for desired topics and subjects.

As It Happens
http://radio.cbc.ca/programs/asithappens/

This Morning
http://radio.cbc.ca/programs/thismorning/

The Nature of Things with David Suzuki
http://www.tv.cbc.ca/nature_of_things/

Out Front (Toronto)
http://radio.cbc.ca/insite/OUT_FRONT
_TORONTO/1999/9/7.html

Quirks and Quarks
http://www.radio.cbc.ca/programs/quirks/index.html

Other special audio and radio shows for the environment include:

The Bioneers
http://newdimensions.org/html/bioneers.html

Deep Ecology for the 21st Century
http://newdimensions.org/html/ecology.html

Earthwatch Radio
http://www.seagrant.wisc.edu/Earthwatch/

GreenWave
http://www.greenradio.com/

The Florida Environment
http://www.floridaenvironment.com/

NatureWatch
http://www.naturewatch.com/

Future Watch
http://www.futurewatchonline.org/

Trash Talk
http://www.trashtalk.org/

Specialty Resources

Environmental Hazards Management Institute (EHMI, http://www.ehmi.org/) presently offers twenty-seven different educational products, all targeted at providing environmental information to various stakeholders. Their unique Environmental Wheels are among the most widely distributed environment-oriented promotional materials distributed. They may be personalized with company, organization, or group name, logo, or message. Topics for which these creative and useful wheels have been developed include: Household Chemical Product Management, Recycling, Home Composting, Energy Conservation, Water Quality and Conservation, Pollution Prevention, Lead Poisoning Prevention, Emergency Preparedness, and Indoor Air Quality. If a student is conducting a community service project, event, or activity requiring distribution of useful, informative, and educational resources, it is suggested that students and teachers contact appropriate local businesses (utilities, hardware stores, grocery stores, nursery, manufacturer) or local government agencies that might be willing to help purchase these information tools in bulk quantities.

Magazines and Journals

The major body of knowledge about the environment is found in technical, subject-specific literature that is well beyond the scope of most school students and many undergraduate college students as well. A good resource for gaining access to this body of scientific and technical environmental literature can be found in the Web site of the National Library for the Environment at http:\\www.cnie.org/Journals.html or http://ublib.buffalo.edu/libraries/units/sel/environ/e-journals.html.

Nonprofessionals can rely on other sources of periodical information. Some general interest or popular news magazines have sections or columns devoted solely to environmental issues. Some of the more familiar ones frequently found in school or public libraries include the following: *Life, Newsweek, Rolling Stone, Time,* and *U.S. News and World Reports.* These popular magazines provide a general, nontechnical overview of a subject as interpreted by science writers rather than by research scientists and may refer to but rarely will cite primary sources of information.

The following limited list of science magazines and trade journals feature articles written for multidisciplinary or general interest audiences and provide significant news, issue reviews, and featured articles on environmental topics and concerns. Articles published in these periodicals are more focused than those in popular magazine but are still interdisciplinary in nature. Some are oriented to amateur scientists and hobbyists. The following group of periodicals more readily mention, quote, and cite the primary sources of information from which the articles are written.

BioScience
Chemical and Engineering News
Discover
EM - Environmental Manager
The Ecologist
Environment
Environmental Protection
Environmental Science and Technology
Fisheries
Geotimes
IEEE Spectrum
Nature
New Scientist
Occupational Hazards
Pollution Engineering
Popular Science
Science

Science News
Scientific American
Scientific American Presents
Water Engineering and Management
Weatherwise

The final category of periodical literature includes environmental and natural history magazines published or produced by environmental advocacy groups and nonprofit organizations. Several of these periodicals have been published for more than 100 years. These resources are worth scouting out in your local school, public, and college libraries. Ask the librarian for help in locating these and other magazines and periodicals related to nature and the environment.

Alternatives Journal (University at Waterloo)
American Forests (American Forests)
Amicus Journal (Natural Resources Defense Council)
Audubon Magazine (Audubon Society)
BioCycle: Journal of Composting and Recycling (J. G. Press)
Canadian Geographic (Canadian Geographic Society)
Defenders (Defenders of Wildlife)
E—The Environmental Magazine (Earth Action Network)
EDF Letter (Environmental Defense)
Earth Focus (Friends of the Earth)
Earth Island Journal (Earth Island Institute)
Earth Times (Earth Times)
Earth Watch (Earth Watch institute)
The Electronic Green Journal (University of Idaho)
Endangered Species (Thylacine Publishing)
Fauna (Fauna, Inc.)
Grist Magazine (Earth Day Network)

In Business (In Business)
International Wildlife (National Wildlife Federation)
Mother Earth News (Sussex Publishers)
National Geographic (National Geographic Society)
National Parks (National Parks Conservation Association)
National Wildlife (National Wildlife Federation)
Natural History (American Museum of Natural History)
Orion: People and Nature (Orion Society/Myrin Institute)
Orion Afield (Orion Society/Myrin Institute)
Our Planet (United Nations Environment Program)
Rachel's Environment and Health Weekly (Environmental Research Foundation)
Sierra (Sierra Club)
Solar Today (American Solar Energy Society)
Whole Earth (Point Foundation)
Wildlife Conservation (Wildlife Conservation Society)
World Watch (World Watch Institute)

Professional Resources for Librarians and Teachers

Journals

Students, teachers, librarians, and parents can keep abreast of new reference works, Internet resources, and audiovisual media from a number of publications through dedicated columns, reviews, and feature articles in the following professional journals.

School Libraries in Canada: The Journal of the Canadian School Library Association (Ladysmith, BC: Canadian School Library Association)

School Library Journal (New York, NY: Cahners)

School Library Media Activities Monthly (Baltimore, MD: LMS Associates)

SB&F (Science Books and Films) (Washington, DC: American Association for the Advancement of Science)

The Science Teacher (Arlington, VA: National Science Teachers Association)

Teacher Librarian: The Journal of School Library Professionals (Seattle, WA: Rockland Press)

Today's Librarian (Phoenix, AZ: Virgo Publishing)

Librarian: The Video Review Guide for Libraries (Seabeck, WI: Video Librarian)

Professional Associations

The following library associations and affiliates also provide useful resources in their publications, newsletters, and conferences for keeping current with resources.

American Library Association (ALA)
http://www.ala.org/

American Association of School Librarians
http://www.ala.org/aasl/index.html

Canadian Association of Public Libraries
http://www.cla.ca/top/capl/capl.htm

Canadian School Library Association
http://www.cla.ca/divsites/csla/

Public Library Association
http://www.pla.org/

(ALA) Social Responsibilities Round Table
http://libr.org/SRRT/

(ALA) Task Force on the Environment
http://www.ala.org/alaorg/rtables/srrt/tfoe/

(ALA) Young Adult Library Services Association
http://www.ala.org/yalsa/

Agencies and Organizations

There are thousands of government agencies and nongovernment organizations and associations that have responsibility for the management of or the advocacy of environmental quality and conservation and protection of natural resources. They regularly provide fact sheets, background information, brochures, flyers, report summaries, newsletters, journals or magazines, career information, job announcements, educational resources for teachers and students, and other resources relevant to their environmental missions. Government agencies typically have jurisdiction over a specific aspect of the environment or natural resources. In many cases they are responsible for enacting and enforcing environmental laws and regulations. Very often government agencies can be the source of scientific and technical information that result from ongoing research, monitoring, assessment, or measuring activities.

Nongovernment organizations on the other hand represent a rather diverse group of public interest advocacy groups, professional and trade societies or associations, and special interest groups that lobby on behalf of an industry or business. These groups may sponsor research but are more often associated with policy making and decision making about specific environmental, ecological, or natural resources issues. They often provide nontechnical information or summaries of technical data and information. They can be rich sources of information for educators, teachers, and students at all levels in both formal and nonformal education settings. The National Wildlife Federation's Annual Conservation Directory provides the single best resource for identifying environmental organizations worldwide with more detailed listings for the United States and Canada.

International Environmental Sites

The United Nations Environment Programme (UNEP), http://www.unep.org, was formed as a result of the 1972 United Nations Conference on Environment and Development (UNCED). The conference created an international forum for dialogue on international, transboundary, and global environmental problems. INFOTERRA is the UNEP database and network, http://www.unep.org/infoterra; which includes INFOTERRA/USA, http://www.epa.gov/earlink1/INFOTERRA/index.html, which is located and operated by the U.S. Environmental Protection Agency.

The World Health Organization (WHO), http://www.who.int/, is an international organization that supports research and policy developments related to all aspects of health. Environmental health issues are addressed by the WHO Program for the Promotion of Environmental Health, http://www.who.int/peh/ccs/index.html; Intergovernmental Forum on Chemical Safety, http://www.who.int/ifcs/ifcsinfo.htm; and Inter-Organization Program for the Sound Management of Chemicals, http://www.who.int/iomc/

Weather prediction, air pollution research, climate change related activities, ozone layer depletion studies, and tropical storm forecasting, are some of the activities the World Meteorological Organization (WMO), http://www.wmo.ch/, coordinates globally. WMO operates within the United Nations and is a 185-member organization providing a scientific voice on the state and behavior of the earth's atmosphere and climate.

The International Geosphere-Biosphere Programme (IGBP), http://www.igbp.kva.se/, is an interdisciplinary scientific organization established by the International Council for Science (ICSU), http://www.icsu.org/. The program was instituted by ICSU in 1986, and the IGBP Secretariat was established at the Royal Swedish Academy of Sciences in 1987. These organizations provide scientific and technical oversight of international environmental and ecological research efforts and fill critical quality assurance functions. They also provide an outstanding example of the complexities of environmental issues.

U.S. Government Departments and Agencies

The Agency for Toxic Substances and Disease Registry (ATSDR), http://atsdr1.atsdr.cdc.gov:8080/atsdrhome.html, is a division of the Public Health Service. ATSDR's mission is to prevent exposure to and adverse health effects from exposure to hazardous substances from toxic and chemical waste sites, unplanned releases of

toxic chemicals into the environment, and from other exposures to toxic chemicals. There are several major components of ATSDR worth noting. The first is the HazDat Database, http://www.atsdr.cdc.gov/hazdat.html, an online searchable inventory of environmental health data resulting from exposure to toxic materials. Public

Health Statements,
http://www.atsdr.cdc.gov/HAC/PHA/, are
overviews of the toxicological effects of specific
chemicals or classes of chemicals most frequently
found in hazardous waste sites.

The Centers for Disease Control and Prevention
(CDC), http://www.cdc.gov/, is another Division
of the Public Health Service with a broad mandate
to protect the health of the nation from illness and
disease. One of the subunits of CDC is the
National Center for Environmental Health,
http://www.cdc.gov/nceh/ncehhome.htm, which
maintains the Lead Poisoning Prevention
Program, http://www.cdc.gov/nceh/pubcatns
/97fsheet/leadfcts/leadfcts.htm.

The Council on Environmental Quality (CEQ),
http://www.whitehouse.gov/CEQ/, was created by
the National Environmental Policy Act of 1969; see
NEPANet, http://ceq.eh.doe.gov/nepa/nepanet
.htm. CEQ prepares the President's Annual Report
to Congress under the title Environmental Quality.
Online copies are available from 1995 to the
present.

The U.S. Department of Agriculture,
http://www.usda.gov/, was established in 1862 to
improve farming and agriculture. Included in its
mission is the protection of the environment,
including soil, water, forests, and natural resources.
The USDA maintains the National Agricultural
Library, http://www.nalusda.gov/, which maintains
the bibliographic database, AGRICOLA
(AGRICultural OnLine Access), which is available
free at http://www.nalusda.gov/ag98/. AgNIC
(Agriculture Network Information Center),
http://www.agnic.org/, is a major information
component of the USDA for general nontechnical
information. The Cooperative State Research,
Education, and Extension Service,
http://www/reeusda.gov/, is one of the most visible
environmental and natural resources outreach
programs with an extensive network of
cooperative extension agents working at land grant
universities, which operate county extension
services throughout the country. The Water
Quality Information Center,
http://www.nal.usda.gov/wqic/index.html, is one
of several information centers maintained by the
National Agricultural Library. The U.S. Forest
Service, http://www.fs.fed.us/, is a program within
the Department of Agriculture.

The Department of Energy (DOE),
http://home.doe.gov/, is responsible for all
research and development activities associated
with the exploration and development, delivery
and utilization of all energy resources. The DOE
Office of Biological and Environmental Research,
http://www.er.doe.gov/production/ober/about
.html, is responsible for environmental research
including ecological studies and restoration
operations. The Office of Scientific and Technical
Information (OSTI), http://www.osti.gov/, is the
information management division of DOE and
provides three major information delivery
programs: the DOE Information Bridge, a
searchable database of technical information,
http://www.doe.gov/bridge/; the DOE Reports
Bibliographic Database,
http://apollo.osti.gov/html/dra/dra.html; and the
DOE Human Subjects Research Projects Database,
http://www.er.doe.gov/production/oher/humsubj/
database.html. The Library without Walls,
http://lib-www.lanl.gov/lww/welcome.html, is a
new information access initiative in DOE. DOE
National Laboratories and Other Laboratories,
http://home.doe.gov/people/peopnl.htm, is an
extensive network of research and development
facilities distributed across the nation. Each
national laboratory conducts extensive ecological
and environmental research. Their public
information offices are a good place to start when
looking for information.

The Environmental Protection Agency (EPA),
http://www.epa.gov, is the single largest U.S.
federal agency devoted to environmental research,
policy, education, communication, and
information. The agency is divided into several
subject-related offices: Office of Air Resources,
http://www.epa.gov/oar/; Office of Water,
http://www.epa.gov/OW/; and Office of
Prevention, Pesticides, and Toxic Substances,
http://www.epa.gov/internet/oppts/. The EPA also
maintains a number of environmental data and
information sources such as Databases and
Software, http://www.epa.gov/epahome/Data.html;
EPA Clearinghouses, http://www.epa.gov/epahome
/clearing.htm; and EPA Hotlines, http://www.epa
.gov/epahome/hotline.htm. EPA also provides
technical and nontechnical information for
various environmental issues such as the EPA's
Concerned Citizen Resources, http://www.epa
.gov/epahome/citizen.htm; the comprehensive
Environmental Quality Web Site,
http://www.epa.gov/ceis/; the National Lead
Information Center, http://www.epa.gov
/opptintr/lead/nlic.htm; the Office of
Environmental Justice, http://es.epa.gov/oeca/oej/;
and Act Locally! A Catalog of Tools for
Community-Based Work, http://www.epa.gov
/opptintr/cbep/actlocal/.

The U.S. Environmental Protection Agency (EPA) is divided into 10 geographic Regions, each with a central Regional Office. Each EPA Regional Office is responsible within selected states for executing agency programs, considering regional needs, and implementing federal environmental laws. A searchable map and text-links are provided for the EPA Regional Offices at http://www.epa.gov/epahome/locate2.htm. EnviroMapper, http://www.epa.gov/enviro/html/em/index.html, is a dynamic service developed by the EPA to view and query subnational environmental information, including drinking water, toxic and air releases, hazardous waste, water discharge permits, and Superfund sites. SurfYourWatershed, http://www.epa.gov/surf/, is another EPA cartographic resource to locate, use, and share information related to watersheds in the U.S. Dynamic links to various toxic releases in specified watersheds and information on hazardous waste sites in these watersheds are found in this resource. EPA's EnviroFacts, http://www.epa.gov /enviro/index_java.html, provides state and community-level data and information for the following topics: air releases, toxic releases, water discharge permits, toxic release inventory facilities, risk management information, safe drinking water, brownfields, microbial and disinfection by-products, and Superfund sites.

One of the most important and effective databases developed in the past twenty years is the EPA's Toxic Release Inventory (TRI), a searchable database of releases of chemicals into the environment by specific chemicals and specific business and industry categories. Details on TRI are found at http://www.epa.gov/opptintr /tri/index.html with several other sites available to search TRI data: Accessing and Using TRI Data http://www.epa.gov/opptintr/tri/access.htm, and the Office of Solid Waste and Emergency Response, http://www.epa.gov/swerrims/.

The National Institutes of Health (NIH), http://www.nih.gov/, supports biomedical research. NIH maintains the National Library of Medicine, http://www.nlm.nih.gov/, as the nation's primary medical information resource. Included in NLM's arsenal of information resources is MEDLINE, a free bibliographic database. Other information resources include Toxicology and Environmental Health Information, http://sis.nlm.nih.gov/tehip.htm; Toxicology Databases, http://sis.nlm.nih.gov /ToxSearch.htm; Chemical Information,

http://chem.sis.nlm.nih.gov/chemindex.html; ToxNet, http://sis.nlm.nih.gov/sis1/; and ToxLine, http://igm.nlm.nih.gov/cgi-bin/doler?account =++&password=++&datafile=toxline

The National Institute of Environmental Health Sciences, http://www.niehs.nih.gov/, is another NIH institute providing specific support for environmental health issues. NIEHS maintains Community Outreach and Centers, http://www.niehs.nih.gov/external/outreach.htm, and Community Outreach and Education Programs, http://www.niehs.nih.gov/centers/coep/coepcver .htm, at all NIEHS university-based Environmental Health Science Centers, http://www.niehs.nih.gov/centers/, many of which sponsor summer science research programs for high school students and teachers. The NIEHS Office of Communications and Public Liaison provides online access to a very useful book for students, *Environmental Diseases from A to Z*, http://www.niehs.nih.gov/external/a2z/home.htm. A Teachers Support site, http://www.niehs.nih.gov/external/teacher.htm, is a very useful resource for teachers and students wishing to explore the issues of environmental health more closely, including Curriculum Materials targeted to specific grade levels, K-2, 3-5, 6-8, and 9-12.

Another NIH institute is the National Institute for Occupational Safety and Health, http://www.cdc.gov/niosh/homepage.html, which provides research support for occupational medicine and related safety and health. NIOSH Databases, http://www.cdc.gov/niosh/database.html, can be searched at no cost, including the NIOSHTIC Bibliographic Database, http://www.cdc.gov/niosh/database.html, the NIOSH Pocket Guide to Chemical Hazards, http://www.cdc.gov/niosh/npg/npg.html, and RTECS, the Registry of Toxic Effects of Chemical Substances, http://www.cdc.gov/niosh/rtecs.html.

The National Aeronautics and Space Administration, http://www.nasa.gov, does not come readily to mind when thinking abut the environment, however, NASA offers a global look at our natural environment. NASA's Earth Observatory, http://earthobservatory.nasa.gov/, is the first place to go for a virtual tour of the earth and its resources. NASA also has the Destination Earth site, http://www.earth.nasa.gov/, another trove of information. For out of the world experiences go to the Exobiology Web site, http://exobiology.nasa.gov/. NASA supports the Global Change Master Directory, http://gcmd.nasa.gov/, a gateway search engine for data and information about global climate change. Another NASA resource for global change information is the Earth Observing System, http://ltpwww.gsfc.nasa.gov/eospso/web/htdocs/. Regardless of where you go or how you get there, exploring the NASA site for environmental information will be well worth the time and effort.

The National Oceanic and Atmospheric Administration (NOAA), http://www.noaa.gov/, was created in 1970 as the result of a presidential governmental reorganization. NOAA is responsible for atmospheric and marine resources. Environmental information resources at NOAA, http://www.noaa.gov/env_info.htm, include the

National Environmental Data Index (NEDI), http://www.nedi.gov/, and its catalog, http://www.nedi.gov/NEDI-Catalog/. The National Sea Grant program, http://www.nsgo.seagrant.org/, is a comprehensive network among those states with borders on the Pacific and Atlantic Oceans, the Gulf of Mexico, and the Great Lakes. See also the Sea Grant extension services through the National Sea Grant College Program, http://www.nsgo.seagrant.org/NationalSeaGrant.html

The Department of Interior, http://www.doi.gov, is the agency given stewardship over natural resources. Within the Department of Interior you will find the Fish and Wildlife Service, http://www.fws.gov/, the U.S. Geological Survey, http://www.usgs.gov/, and the National Parks Service, http://www.nps.gov/. The National Biological Information Infrastructure, http://www.nbii.gov/index.html, is part of the USGS.

State Environmental Agencies

It is beyond the limited scope of this overview of reference sources to provide the individual state Web sites for state agencies, departments, offices, and bureaus having jurisdiction over environmental quality and health, natural resources, and fish and game concerns. There are, fortunately, several outstanding sources for identifying these state agencies.

Yahoo is one of the most widely used Internet Directory search engines. Individual state information is obtained by going to the Yahoo "United States" site at http://dir.yahoo.com/Regional/U_S_States/. From this site one simply selects the state desired (scroll to state resources sought), or one can link to the inventory of "Counties and Regions" (or "Parishes and Regions"), Metropolitan Area, or Cities. Additional state resources may also be listed in sections of a Yahoo "states" search in categories such as "Community and Culture," "Government," "Health," or "Science."

Piper Resources provides a metasite gateway service for locating state and local government information on the Internet. Their "State and Local Government on the Net" site, http://www.piperinfo.com/state/index.cfm, provides a frequently updated directory of links to state and

local government sponsored and controlled resources on the Internet. For each state, information is provided for the state's official Web site, http://www.state.[state's two-letter postal abbreviation].us (for example, New York's site is http://www.state.ny.us). Also available are links to state offices (governor, attorney general, secretary of state); legislative, judicial, and executive branches (state agencies, departments, and bureaus are typically listed under the state's executive branch); special state boards and commissions; links to individual county and city Web sites; state library networks; and other information (state government associations, leagues and organizations).

The Library of Congress (LOC) provides a meta-site gateway for individual state governments at http://lcweb.loc.gov/global/state/stategov.html. From this LOC site comprehensive access is provided from a "Meta-Indexes for State and Local Government Information" and links to individual LOC state pages via the LOC "State Government Information Site.

The U.S. Fish and Wildlife Service (FWS) provides a rather comprehensive state-by-state listing of state fish and wildlife, conservation, and natural resources agencies, as well as FWS state offices at http://offices.fws.gov/statelinks.html.

Nonprofit Public Interest Organizations

There are hundreds of nonprofit organizations that deal with various aspects of the environment, ecology, and natural resources. The following list represents the breadth of coverage.

Center for Health, Environment, and Justice (formerly the Citizens' Clearinghouse for Hazardous Wastes) http://www.essential.org/cchw/

Defenders of Wildlife
http://www.defenders.org/

Earthwatch Institute
http://www.earthwatch.org/

Environmental Defense Fund's Chemical Scorecard
http://www.scorecard.org/

Greenpeace
http://www.greenpeace.org/

National Audubon Society
http://www.audubon.org/

National Environmental Health Association
http://www.neha.org/

National Council for Science and the Environment
(formerly the Committee for the National Institute
for the Environment)
http://www.cnie.org

National Safety Council–Environmental Health Center
http://www.nsc.org/ehc.htm

National Wildlife Federation
http://www.nwf.org/

Natural Resources Defense Council
http://www.nrdc.org/

The Nature Conservancy
http://www.tnc.org/

Pollution Maps of the United States and
the World
http://www.mapcruzin.com/global_toxmaps.htm

Public Interest Research Groups (States)
http://www.pirg.org/

Rachel's Environment and Health Weekly
(Environmental Research Foundation)
http://www.rachel.org/home_eng.htm

Sierra Club
http://www.sierraclub.org/

Wildlife Society
http://www.wildlife.org/

Worldwatch Institute
http://www.worldwatch.org/

Professional Societies and Associations

The Web sites of professional societies and
associations contain a wealth of information
including newsletters, trade magazines, news
briefs, reports, events, meetings, career
counseling and job openings, and continuing
education opportunities. Some of these
organizations support high school chapters and
can provide speakers for classes and may even
arrange special field trips. Check the sites for
local or regional chapters. Following is a small
representative sampling among the scores of
such groups.

Air and Waste Management Association
http://www.awma.org/

American Academy of Environmental Engineers
http://www.enviro-engrs.org/

American Chemical Society
http://www.acs.org/

American Fisheries Society
http://www.fisheries.org/

American Geophysical Union
http://www.agu.org/

American Water Resources Association
http://www.awra.org/

Canadian Geotechnical Society
http://www.cgs.ca/

Canadian Water Resources Association
http://www.cwra.org/

Ecological Society of America
http://esa.sdsc.edu/

Environmental and Engineering Geophysical Society
http://www.eegs.org/

Environmental Council of the States
http://www.sso.org/ecos/

International Society for Environmental Ethics
http://www.cep.unt.edu/ISEE.html

National Association of Environmental
Professionals
http://www.enfo.com/NAEP/

National Environmental Health Association
http://www.neha.org/

Physicians for Social Responsibility—Environment and Health Program
http://www.psr.org/enviro.htm

Society for Ecological Restoration
http://www.ser.org/

Society of American Foresters
http://www.safnet.org/

Society of Environmental Journalism
http://www.sej.org/

Society of Environmental Toxicology and Chemistry
http://www.setac.org

Society of Toxicology
http://www.toxicology.org/

Canadian Government Sites

Like the United States, Canada has an extensive network of government information at the national and provincial levels. A convenient starting place is the Yahoo search engine at http://dir.yahoo.com/Regional/Countries/Canada/. Look under categories Business and Economy, Science, Society and Culture, and Government (Departments).

Environment
http://dir.yahoo.com/Regional/Countries/Canada/Government/Departments/Environment/

Fisheries and Oceans
http://dir.yahoo.com/Regional/Countries/Canada/Government/Departments/Fisheries_and_Oceans/

Natural Resources
http://dir.yahoo.com/Regional/Countries/Canada/Government/Departments/Natural_Resources/

Parks
http://www.parkscanada.gc.ca/

Government listings for Canadian Provinces can be located with Yahoo at http://dir.yahoo.com/Regional/Countries/Canada/Provinces_and_Territories/

Places to Visit

Mention the word environment and one of the first images that comes to mind is the outdoors—our ambient environment. It is not surprising then for people to want to have places to go to experience the environment firsthand. Where can you go to study or celebrate the environment? Nearly every nation has developed a network of national parks that display the natural beauty and resources of that country. These areas have been set aside to protect those resources and facilitate the sharing of the resources they hold. Around the world countries have established more than 3,500 national parks and protected areas.

National Parks and Related Institutions

The U.S. National Park Service, http://www.nps.gov, is the steward over the entire system of more than 300 National Parks, National Preserves, Wilderness Areas, National Rivers, National Scenic Trails, National Scenic and Wild Rivers, National Lakeshores, Wildlife Refuges, National Seashores, and Recreation Areas in the United States (http://www.nps.gov/legacy/nomenclature.html). In addition to the scenic beauty and recreational potential of these preserves are the rich educational opportunities for students and teachers. Each park has its own unique programs, which are outlined in the Parks Guide, http://www.nps.gov/parks.html. Opportunities range from one-day field trips to summer employment or research positions for high school students and their teachers.

Similarly, the U.S. Forest Service manages the national network of National Forests and National Grasslands, http://www.fs.fed.us/links/forests.shtml.

The National Oceanic and Atmospheric Administration has jurisdiction over the National Marine Sanctuaries, http://www.sanctuaries.nos.noaa.gov/. All of the U.S. National sites are identified in the National Wildlife Federation's *Conservation Directory* also.

Canada maintains similar programs of National Parks, http://dir.yahoo.com/Regional/Countries/Canada/Recreation_and_Sports/Outdoors/Parks_and_Public_Lands/National_Parks/. See also National Forests, http://nfis.cfs.nrcan.gc.ca/.

Each U.S. state also maintains a network of state parks. Methods for identifying state park agencies are outlined below. In general state parks are geared more toward recreational use and may not have extensive research programs conducted in them. It is best to check each state's agency that manages state parks for particular aspects of research and study that can be conducted.

Research Programs

An often overlooked resource for places to go is the network of ecological research sites operating in conjunction with the Man and the Biosphere (MAB) Program, http://www.usmab.org/home2.html. The mission of the U.S. MAB Program is to "explore, demonstrate, promote, and encourage harmonious relationships between people and their environments building on the MAB network of Biosphere Reserves and interdisciplinary research" within the international MAB network. The U.S. MAB program has nominated 47 biosphere reserves that are areas of special significance because of their research, educational, or resource management programs. A list of each biosphere reserve in the U.S. MAB is found at http://www.usmab.org/brprogram/usbrl.html. The directory listings of these resources are found online at http://www.usmab.org/misc/us contact.html. The full international inventory of biopreserves in the MAB Program is found online at http://www.unesco.org/mab/bios1-2.htm, including the ten biopreserves located in Canada, http://www.unesco.org/mab/br/brdir/europe-n/canada.htm.

In recent years the word *ecotourism* crept into our environmental lexicon. The concept of ecotourism is to take the dream vacation of a lifetime in a pristine, little visited part of the world. Earthwatch Institute, http://www.earthwatch.org, was the first organization to engage the public at large in ongoing scientific field research projects. In 2000 Earthwatch supported more than 130 expeditions in 48 countries and more than 2,000 people, many of them high school aged, joined ranks with 200 scientists to collect ecological and environmental data and become part of research campaigns and expeditions. Programs emphasize research, conservation, and education and will address specific issues (e.g., endangered ecosystems, biodiversity, global change, cultural diversity, and world health).

Museums, Science Centers, and Community Programs

Museums and science and nature centers are other institutions where studying the environment and natural resources is possible. Larger, metropolitan science museums cover a broad disciplinary array of exhibits, including those related to natural resources and the environment. The American Museum of Natural History, http://www.amnh.org/; the Academy of Natural Sciences of Philadelphia, http://www.acnatsci.org/; Carnegie Museum of Natural History, the Carnegie Museum of Natural History, http://www.clpgh.org/cmnh/; San Francisco's Exploratorium, http://www.exploratorium.edu/; Toronto's Ontario Science Center, http://www.osc.on.ca/, and the Smithsonian Institution's (http://www.si.edu/) National Museum of Natural History, http://www.mnh.si.edu/, and Environmental Research Center, http://www.serc.si.edu/, serve as examples of science museums. Some museums support remote facilities or nature centers where more intense attention is devoted to the ambient environment. These facilities often provide rich settings where young scientists and environmentalists can work on special projects and programs. Students and teachers alike will find many educational opportunities in science museums near, far, and on the Internet.

Places to look for museums of science and natural history include the standard Yellow Pages of the local phone book. Museums also support impressive Web sites, which can be identified from any number of search engines and metasites, such as MUSEE, one of the most comprehensive of museum Web inventories, http://www.musee-online.org/frames2.htm; the University of Washington's Science Museums and Exhibits site, http://www-hpcc.astro.washington.edu/scied/museum.html; and Yahoo's List of Science Museums and Exhibits, http://dir.yahoo.com/science/museums_and_exhibits/.

Zoos, aquariums, arboretums, and botanical gardens are other places worth exploring for extending and enhancing the environmental experience. Exotic flora and fauna from the most distant parts of the world are aggregated in these special collections of living resources. As with museums and nature centers, zoos and botanical gardens provide education programs for teachers and students. These range from one-hour presentations to class visits and from weekend classes to weeklong or summer programs of more epic dimensions. Consult the MUSEE site, http://www.musee-online.org/frames2.htm.

The first place you might want to investigate environmental issues and research is in your own backyard. Literally speaking, your backyard, whether it be an urban apartment complex (and a parking lot is your backyard) or a rural home surrounded by hundreds or thousands of acres of land, is a living ecosystem and is waiting to be explored.

More broadly speaking your backyard represents the neighborhood or community in which you live. Another word that has entered or environmental vocabulary in recent years is *sustainability*. There are increasing needs and desires among various parts of our communities to seek a balance in the way our urban, suburban, and rural communities develop and grow. The American Library Association launched a major Libraries Build Sustainable Communities program in 1999 to allow libraries and librarians to become integral components of the community decision-making process. Included in this effort was the development and identification of resources to help communities find the means to achieve a balance of ecology, economics, and equity. ALA's LBSC project can be found at http://www.ala.org/sustainablecommunities.

Thematic Indexes

Page numbers in **boldface type** indicate full articles on a subject. Page numbers in *italic type* indicate illustrations or other graphics. All numbers preceding colons indicate volume numbers.

Page numbers in **boldface type** indicate full articles on a subject. Page numbers in *italic type* indicate illustrations or other graphics. All numbers preceding colons indicate volume numbers.

sources and sinks **2:**158
tropical rain forest **10:**1270–1271
water cycle and **10:**1359
See also photosynthesis
carbon dioxide
atmospheric **1:**84–87, **2:**152, 153, 157, 158–159, 169, 185, 187
in carbon cycle **2:**157–159, 187
as coal combustion by-product **2:**194, 197, 199
deforestation and **2:**254, **4:**525, 549, **5:**617
as greenhouse gas **2:**185, 188, **5:**612, 615, 616
as natural gas product **7:**861, 862
oxygen cycle and **7:**929, 930, 932, *933*, 937
percent in earth's atmosphere **5:**573
petroleum emissions **8:**982, 983
sources, storage, and level **1:**84–85
carbon monoxide **2:160–161**
air pollution **1:**18, 19, *20*, 21, 22, *23*, **2:**160
Clean Air Act **2:**178
emissions **2:**160, 178, **8:**982
emissions control **1:**92
poisoning symptoms **2:**161
carbonyl sulfide **1:**86
carnivores **3:**316, 361, **4:**516, 522
carotenoids **3:**278
carrying capacity **2:164–165**, 217, **3:**331, **8:**1008–1015
cat **1:**35
caterpillar **6:***717*
cellulose **9:**1185
chain reaction, nuclear **7:**893–894
chaparral **1:**119, *123*, **2:170–171**, **8:**1053
charcoal **2:**153, 173
chemical energy. *See* energy, chemical
chemolithotrophic bacteria **1:**94, 96, 97
chicken **1:**35
chimpanzee **1:***48*, 49, 50, 51
chitin **6:**711, 712
chlorine **2:**160, 185, **10:**1368, 1386
ozone layer **7:**938–939, 940
PCB **7:**959
chlorophyll **1:**24, **8:**986, 988, *989*
chloroplasts **3:**292, **8:**987, *988*, *989*
chrysotile **1:**76
classic static stability **3:**320, 321, 322, 324, 343
climate change **2:184–189**
agriculture and **1:**63–64
algae function **1:**24–25
from Aral Sea diminishment **1:**60–61
archeological studies **1:**63, 64
architecture and **1:**68–69
as biodiversity threat **1:**114
as biome determinant **1:**120
carbon cycle and **3:**310, 350
carbon dioxide level and **1:**86, **2:**154, 157, 158–159
coral reef effects **2:**226–227, 229
desertification and **3:**271
Earth Observing System data **3:**310, 311
ecologism and **4:**438
El Niño **3:**356–357, **7:**912
environmental journalism on **6:**737
forests and **4:**523, 525, 526–527, 549
glacier mass balance **5:**592–594, 597
global warming **6:**705
health effects **5:**635

human activity and **8:**1107
industrial ecology and **6:**699
international studies of **6:**718, 719
Lake Baikal effects from **1:**100
museum exhibit **7:***848*
oceanic heat storage and **3:**311
paleontological study **7:**943
petroleum combustion and **8:**982, 983
Pollution Probe campaign **8:**1005
quality of life and **8:**1040
sustainable development **9:**1177
termite methane emission and **9:**1205
tourism and **9:**1248, 1249–1250
tropical rain forest destruction and **10:**1270–1272
tundra sensitivity to **10:**1281, 1283
UN conferences and convention **4:**428, **10:**1292, 1293, 1297
water cycle and **10:**1353, 1357, 1359
Clostridium **1:**95, **3:**279
clouds
climate change **2:**184, 185–186
Earth Observing System **3:**309
heat energy **3:**394
meteorology **6:**817
water cycle **10:**1356, *1358*
coal **2:194–199**
air pollution **1:**18, *22*
ash generation **1:**78
carbon in **2:**152, 155, 157, 158, 194, 197
cogeneration techniques **2:***202*
combustion-releasing acid rain **1:**2, 3
electricity generation **3:***348*, 349, 376, 388
as energy source **3:**376, 377, 378
as fossil fuel **5:**558, 559
gasification **5:**562, *563*
historical use of **9:**1191–1192
industrialization and **6:**700–703, 705
strip mining **2:***153*, 213
coastline. *See* shoreline
cockroach **9:**1184–1185
combustion
carbon monoxide **2:**160, 161
cogeneration **2:**202
nitrogen and **7:**882, 888
oxygen and **7:**930
unvented **1:**21
waste materials **10:**1339–1340, 1342, 1344–1345
common descent theory **4:**476–477
coniferous forest. *See* forest, coniferous
continental drift theory **7:**907
continental ice sheets **2:**186–187, **5:**592, 594, 597
continental shelves **7:**910, 911
coral reef **2:224–229**, **7:***936*
biodiversity **1:**114
biome **1:***122*
Great Barrier Reef **5:**604, 605
keystone species **6:**740, 741
as natural habitat 7, 867
oil spill effects **7:**914
symbiosis **2:**225, 226
cosmic radiation **8:**1043–1044
covalent bond **7:**880, 972
Cromwell current **7:**911
crude oils. *See* petroleum
currents, ocean **7:**911

cyanobacteria **1:**94, *97*, **7:***934*
cyclone **1:***83*, **7:**857

deciduous forest. *See* forest, deciduous
deer **1:**35
deer ticks **3:**289
denitrification **1:**97, **7:**885–886, 888
desalination **2:258–259**, **7:**924
desert **2:260–265**
climate change and **2:**187, **8:**1107
grassland **2:**260, 264, **3:**270, **5:**599, 603
as habitat **7:***867*, *869*
Operation Desert Bloom **2:***262*, **7:***924*–925
desertification **3:270–273**, **4:***504*, **5:**673
causes of **3:**273, 300
climate change and **8:**1107
desert and **2:**260–261
drought and **3:**300, 301
international agreements **4:**430–431
irrigation effects **3:**272–273
rangeland **8:**1051, 1053
soil conservation **9:**1149
soil erosion **9:**1157
sustainable development **9:**1178
water and **10:**1353
deuterium **7:**894
dewdrops **7:**867
diamonds **2:**152, 156
dieldrin **5:**606
dinosaurs **7:***942*
disaster. *See* natural disaster
dissolved oxygen **3:290–291**
lake **6:**746–747
DNA **3:**292–293, **5:**578–579, **7:**881
carcinogenic effects on **2:**146
chemical mutation **9:**1230–1231
cloning **2:**190, 191, 192, 193
code **5:**653
evolution and **4:**475
food irradiation and **6:**723, 724
genetic code **2:**133–134
Human Genome Project **5:**652–655
phosphorus and **8:**984
plant domestication analysis **8:**997
recombinant technology **2:**132–133
structure **5:**577, 653
viral **10:**1318
dodo (bird) **1:**114
dog **1:**32–33
donkey **1:**34
drainage area. *See* watershed
drinking water **3:298–299**, **9:**1164, **10:**1354
acid rain and **1:**3
contaminants **1:**99, **3:**277, 298, 299, **5:**631–632, 634, **10:**1364, 1367
corrosion of **1:**3
Danube River basin **2:**236
deep well injection **2:**251
desalination **2:**258, 259
groundwater as source **5:**622
nitrate content **7:**882, 889
quality controls **2:**180, 181
radon contamination **8:**1047
safety engineering **10:**1386
scarcity **9:**1127
dromedary **1:**34
drought **1:**106, **2:**158, **3:300–301**, **7:***855*
agricultural **1:**11

Page numbers in **boldface type** indicate full articles on a subject. Page numbers in *italic type* indicate illustrations or other graphics. All numbers preceding colons indicate volume numbers.

Page numbers in **boldface type** indicate full articles on a subject. Page numbers in *italic type* indicate illustrations or other graphics. All numbers preceding colons indicate volume numbers.

lithosphere **7:**928, 934–935
Little Ice Age **2:**188, **5:**594–595
littoral zone **9:**1128–1129, *1130*
lizards **2:***170*, 171
llama **1:**34
llanos **1:**119
loam **1:***106*

macaque **1:**49
magma **3:**391
magnetic fields **3:**353
manganese nodules **7:***907*
mangrove **3:***346*, **4:**470, **9:**1128, 1132
maquis **1:**119, 122
marine biome **1:**118–119, *122*
marine creatures. *See* aquatic life
Mars **7:**926–927, 929
marsupials **2:**130, *131*
mercury **2:**179, **6:***696*, **814–815**
    fish contamination **1:**3
    food contamination **3:**277, 281
    toxicity **9:**1224–1226
mesosphere **1:**82, *83*
mesothelioma **1:**77, **2:**174
messenger RNA **3:**292–293
methane **5:**558, **9:**1205
    alternative fuel **5:**556, 564
    bioenergy **1:**115
    carbon cycle **2:**159
    CFC **2:**169
    climate change and **2:**184
    Earth Observing System **3:**309
    greenhouse effect **4:**612, **5:***616*
    natural gas and **7:**858, 859, 862
    overview **7:**861
    petroleum and **7:**972, **8:**983
    trapping efforts **7:**863
methanol **1:**92, 116, **5:**556, 562, 564
methyl mercury **9:**1225–1226
midocean ridges **5:**585
mimicry **10:**1323
mineral cycle **6:828–833**
mineralization, definition of **7:**885
minerals **3:**278, 299, 327–328
    public lands **2:**140
mitochondria **3:**292
molds **5:**570–571
molecule **1:**113
    carbon **2:**154
    CFC **2:**169
    hydrocarbon **7:**972
    nitrogen **7:***881*
    oxygen **7:**932, 934, 936, 937, 938
    toxicology **9:**1231
monomer **8:**1001, 1002
Moon **7:**926, *927*
mosquito **3:**285–286, **6:**710, 717, 786, 787,
    7:946, 960, 961, **10:**1321
mouse **4:**479
mule **1:**34
mushroom **5:**570, 571
mutation **3:**406, **5:**578, 630, **9:**1230–1231
    as adaptation **1:**7
mutualism **3:**331
    tropical rain forest **10:**1268–1269
mycotoxins **3:**280

nanoplankton **1:**84
naphthalene **7:**972

natural balance. *See* ecological stability
natural disaster **7:854–857**, 910, 913
    avalanche **5:**595–596
    from El Niño **3:**356–357
    See also flood; hurricane; volcanic eruption
natural gas **3:**377, 378, **5:**564, **7:858–863**
    as automotive alternative fuel **1:**91–92
    carbon in **2:**152, 153, 157
    cogeneration techniques **2:**202
    as fossil fuel **5:**558–559
    industrialization and **6:**705
    liquefication **5:**556, 560–561, **7:**858,
        859–860, 861, 863
    petrochemicals **2:**173
    petroleum and **7:**972, 974, **8:**983
    synthetic **5:**562
natural habitat **7:864–869**
    biodiversity and **1:**114
    biotic adaptability **2:**131
    of birds **2:**137
    coral reefs **2:**224
    dam building effects **2:**233–235
    ecological change and **3:**324–325
    Elton's animal studies of **3:**360, 361
    endangerment from loss of **3:**341, 371,
        372, 373
    grassland **7:**864, 868, 869
    human impact on **1:**114
    keystone species **6:**740–743
    niche **7:**875–877
    oil spill hazard **7:**915
    preservation **8:**1037
    river and stream **8:**1098–1099
    tropical rain forest **10:**1264
    water availability **10:**1359
    waterfront development effects **10:***1360*
    wetland protection **10:**1368, 1372–1373,
        1376
    wildlife refuges **7:**850–853
natural selection **3:**331, 406, **4:**473, 474,
    10:1333
nebula **3:***355*
neutron **7:**893
niche **7:875–877**
nitrate/nitrite **7:**880–885, 889
    level in water **10:**1367
nitrogen **2:**197, 228, **7:**859, **880–882**
    as basic element **3:**354–355
    carbon and **2:**153, 154, 155
    compost ratio **2:**210, **10:**1339
    distribution of **7:***884*
    emissions **9:**1249, 1250
    fertilizer **4:**486, 487, **5:**620, **7:**881, *885*,
        8:1108, **9:**1192
    liquid 556
    oxygen and **7:**930
    percent in earth's atmosphere **5:**573
    water quality **10:**1367
nitrogen cycle **3:**328, 343, **7:**880–881,
    883–889
    acid rain formation **1:**2, 3–4, 87
    algae and **1:**24–25
    aquaculture effects **1:**57
    in atmosphere **1:**82, 84, 87
    bacteria and **1:**96–97
    dead zone **2:**242, 243, 244, 245
    eutrophication **3:**291, **4:**466
    fungus **5:**570
    indoor air pollution **1:**21, 22, *23*, 223

industrial ecology **6:**696
legume-rhizobia symbiosis **9:**1182–1184
wetland action **10:**1374
nitrogen dioxide **2:**197, **7:**880, 929, 931
nitrogen fixation **7:**886–888
nitrogen oxide **2:**159, 169, **4:**452
    Clean Air Act **2:**178
    coal and **2:**199
    combustion-produced **10:**1339, 1340
    emissions control **1:**92
    nitrogen cycle **7:**880, 885, 888, 889
    oxygen and **7:**929, 930
    petroleum emissions **8:**982
northern spotted owl **4:**533, 546
nucleic acids **7:**881
nutrient cycles. *See* mineral cycle

ocean
    carbon cycle **2:**158
    currents **7:**911
    Earth Observing System data **3:**311
    ecological pyramid **3:**317, 318
    effects on atmosphere **3:**186
    El Niño effects **3:**356–357
    as energy source **3:**395, **8:**1080, *1081*,
        1084, 1085
    fishing industry **4:**498–501
    greenhouse effect **5:**613, 616, 617
    international measures **4:**428–429
    inter-ocean canal effects **2:**145
    layers **7:**911
    oil deposits **7:**910, 974
    oil spill **7:**914, 915
    photosynthesis **8:**988
    seawater farms **7:**924–925
    shoreline **9:**1132
    tidal energy **3:**395, **8:**1080, *1081*, 1085
    UN Law of the Sea Conference
        10:1291–1292
    water cycle **10:**1353, 1355, 1356, 1357
    zones **9:**1128–1131
    See also aquatic life
oil. *See* petroleum
oil shale **5:**562–563
oligotrophic lake **4:**466, 467
oocyte **2:**191, 192
orangutan **1:**49
outer space **7:926–927**
    cosmic radiation **8:**1043–1044
    remote sensing **8:**1076–1079
oxidation **7:**936
oxides **7:**935
oxygen **2:**160, **7:928–931**, 932
    as basic element **3:**354–355
    dissolved oxygen meter **1:***111*
    fuel cell **5:**564
    ozone formation **7:**938, 975
    percent in earth's atmosphere **5:**573, 575
    in water **10:**1350, 1352, 1365–1367
oxygen cycle **7:932–937**
    algae and **1:**24–25
    in atmosphere **1:**82
    carbon and **2:**153–155, 187, **7:**930, 933
    dead zone **2:**242–245
    dissolved oxygen **3:**290–291, **6:**746–747
    in ecosystem **3:**343
    eutrophication **8:**985
    industrial ecology **6:**696
    lake **6:**746–747

Page numbers in **boldface type** indicate full articles on a subject. Page numbers in *italic type* indicate illustrations or other graphics. All numbers preceding colons indicate volume numbers.

Page numbers in **boldface type** indicate full articles on a subject. Page numbers in *italic type* indicate illustrations or other graphics. All numbers preceding colons indicate volume numbers.

Cultural Ecosystems

Page numbers in **boldface type** indicate full articles on a subject. Page numbers in *italic type* indicate illustrations or other graphics. All numbers preceding colons indicate volume numbers.

Page numbers in **boldface type** indicate full articles on a subject. Page numbers in *italic type* indicate illustrations or other graphics. All numbers preceding colons indicate volume numbers.

Page numbers in **boldface type** indicate full articles on a subject. Page numbers in *italic type* indicate illustrations or other graphics. All numbers preceding colons indicate volume numbers.

hybrid plants, **1:**11
hydrocarbons, **1:**84, 92
    Clean Air Act and, **2:**178
    DDT and, **2:**240
    industrialization and, **6:**703–704, 705
    oxygen cycle, **7:**936
    petroleum burning and, **7:**971–975
hydrochloric acid, **2:**172, 173
hydroponics, **5:680–681**
hydropower, **3:**377, 379, **5:682–685, 10:**1354
    Aswan High Dam, **1:**80–81, **5:**639, 682
    electric utility industry, **3:**349, *350,* 351
    environmental law, **4:**433
    glacial source, **5:**597
    irrigation and, **6:***729*
    as renewable energy source, **8:**1080, 1084
    Tennessee Valley Authority, **3:**352, **8:**991,
        9:1202–1203, **10:**1315
    Three Gorges Project, **9:**1200, 1201,
        1216–1219
    tidal, **3:**395
hypnotherapy, **6:**805, 806
illicit drug use, **6:**802–803
immune deficiency disorders, **2:**133
immune system, **2:**163, **10:**1289
    immunotoxicant studies, **9:**1231
    salmonella poisoning vulnerability,
        9:1123
incineration, **5:686–687**
    ash, **1:**78–79
    hazardous waste, **5:**625, 628
    plastic waste, **8:**1002, 1003
    smog, **1:***22*
    technology-forcing regulation,
        9:*1194–1195*
    waste disposal and management, **10:**1333,
        1334, 1335, *1341,* 1343
indoor pollution
    hazardous household waste, **5:**626
    health effects, **5:**631
    Pollution Probe, **8:**1005
    radon, **8:**1044, 1046, 1047
    sick building syndrome, **9:**1136–1137
    tobacco smoke, **1:**18–19, **9:**1222–1223
industrial ecology, **6:696–699,** *719*
    acid rain reduction, **1:***4*
    air pollution, **1:**18–19, *86,* 90–91
    alternative dispute resolution, **1:**30–31
    automotive manufacture, **1:**90–91
    Bhopal toxic gas-leak accident, **1:**108–109
    carbon and, **2:**155
    CFCs and, **2:**168, 169
    chemical industry, **2:**172–175
    clean coal technology, **2:**198–199
    climate change, **2:**188
    ecological literacy, **4:**435
    economic incentive, **3:**332, 337–339
    emission standards, **2:**178–179, 197–198
    green marketing, **5:**610–611
    industrial metabolism, **6:**706–707
    international projects, **6:**718–721
    Lake Baikal, **1:**101
    life cycle analysis, **6:**697, *699,* 770–771
    poor communities, **2:**204–205
    water pollution controls, **2:**180
industrialization, **6:700–705**
    of agriculture, **6:**704
    air pollution from, **1:**18–19, *86,* 90–91
    canal transportation, **2:**145

energy resources, **2:***199*
    fossil fuel and, **5:**558
    geology and, **5:**582
    housing and, **6:**643, 644–645
    megacity development, **6:**808
    nitrogen fixation, **7:**881–882, 888
    noise pollution, **7:**890, 891
    occupational safety, **7:**899
    oxygen uses, **7:**930–931
    smog, **9:**1140, 1141
    technological changes, **9:**1192–1193
    technology-forcing regulation,
        9:1194–1195
    urbanization and, **10:**1306–1307
    utopianism as reaction to, **10:**1317
    water use, **10:**1354
    zoning, **10:**1399
industrial metabolism, **6:706–707**
    toxic metals, **9:**1224
Industrial Revolution. *See* industrialization
industrial waste. *See* hazardous waste,
        industrial
infant mortality rate, **8:**1032
influenza epidemic, **10:**1321
information technology, **6:708–709**
infrared sensors, **8:**1079
injectable contraceptive, **2:**222–223
insecticide, **6:710–711,** 717
    synergistic effect, **9:**1187
insulation, **9:**1209
integrated pest management, **7:**961, 965
Integrated Solid Waste Management,
        10:1342–1343
Intelligent Vehicle/Highway Systems, **8:**1105,
        9:1241, 1244
interferon, **2:**193
internal combustion engine, **6:**702
Internet, **4:**435
interstate highway system, **8:**1102
inventions, **6:**708
iron and steel, **6:**702, 703, **9:**1191, 1193
    steel mills, **6:***700,* 702, 703, 744
irradiation, food, **6:722–727**
    international symbol, **6:***726,* 727
irrigation, **1:***13,* **3:***274,* **4:**502, *504,*
        6:728–732, 823, **10:**1354
    aquifer, **1:**59
    canal, **2:**144–145
    Danube River, **2:**236
    desert, **2:**264
    desertification and, **3:**272–273
    drought and, **3:**300
    ecological effects, **1:**8, 60–61, 62
    grassland, **5:**600, 602, 603
        Green Revolution, **5:**621
    groundwater contamination, **9:**1124
    Ogallala Aquifer, **1:**59
    Operation Desert Bloom, **7:**924–925
    soil salinization, **1:**8, 60, 61, 62
Israeli-Arab wars, **7:**920–921
IUD (intrauterine device), **2:**223, **4:**484

joule, definition of, **7:**888

Kaposi's sarcoma, **2:**146–147
kidney disease, **2:**132
kilowatt units, **3:**349, 352, 385
Korean War, **10:**1325,
        1326–1327

labeling
    ecological, **3:**315–315, *399*
    nutritional, **3:***281*
labor, value of, **2:**150
labor movement, **6:744–745**
    occupational safety and health, **7:**899, 903
landfill, **10:**1333–1338, 1342
    ash-only, **1:**79
    description of process, **10:**1337–1338
    gas recovery, **10:**1339, 1342–1343
    hazardous waste, **5:**627–628
    plastic waste in, **8:**1002–1003
    rangeland, **8:**1054, 1055
Landsat, **8:**1076–1077
landscape architecture, **6:756–761,** **7:**953,
        10:1397
    esthetics, **4:**464–505
    as humanized environment, **5:**657–658
    Olmsted contribution, **7:**916–917, 949
    urban park, **7:**948–951
    ziggurat, **10:**1397
lathe, **9:**1190
lawns, **9:***1160,* 1163
leaching, **9:**1124
lead, **1:***23,* **6:766–767**
    Clean Air Act and, **2:**178
    food contamination, **3:**277, 281
    in gasoline, **6:**703, 766, *767,* **8:**982,
        9:1141
    gastrointestinal effects of, **4:**457
    in hazardous waste, **5:**625
    toxicity, **7:**899, **9:**1224, *1227*
lead acid battery, **5:**555
leukemia, **2:**177, 193
life expectancy, **2:***257,* **5:**630, 662
    global, **8:***1011,* 1013
light pollution, **6:772–773**
light rail transit (LRT), **9:**1246
light water reactors, **7:**894
liquefied natural gas (LNG), **5:**556,
        560–561, **7:**858, 859–860, 861, 863
liver cancer, **2:**174
livestock. *See* animal, domesticated
loom, **9:**1190
lumber, **3:**340–341
lung cancer, **1:**77, **2:**146, 147, *149*
    air pollution and, **1:**21–22
    occupational risk, **7:**900
    radon risk, **8:**1046
    tobacco smoking risk, **9:**1222
lungs. *See* respiratory tract
Lurgi process, **5:**563
Lyme disease, **3:**284, *286,* 288–289, **7:**960
lymph cancers, **2:**146
lymphoma, **2:**146, 147

machinery, farm, **1:**10, 11
machine tools, **9:**1190
macadam roads, **8:**1101
magnetic compass, **9:**1190
malaria, **3:**284, *285,* **6:**717, **786–787,** 799,
        7:945, 946, 947, 961
malignant melanoma, **2:**133, 146
mammogram, **8:***1044*
mapping
    geographic, **5:**580, 581
    geologic, **5:**583, 586–587
marine biology, **7:**905–907
market economy. *See* free market

Page numbers in **boldface type** indicate full articles on a subject. Page numbers in *italic type* indicate illustrations or other graphics. All numbers preceding colons indicate volume numbers.

Page numbers in **boldface type** indicate full articles on a subject. Page numbers in *italic type* indicate illustrations or other graphics. All numbers preceding colons indicate volume numbers.

History of Environmentalism

Page numbers in **boldface type** indicate full articles on a subject. Page numbers in *italic type* indicate illustrations or other graphics. All numbers preceding colons indicate volume numbers.

Sierra Club, **2:**214, **4:**436, 437, 442, 447, 461, 530, **5:**619, **6:**744, 745, **7:**843, **8:**1018, 1034, 1036

Sierra Club Legal Defense Fund, **4:**423, 447, **8:**1034, 1036, 1037

Silent Spring (Carson), **1:**12, **2:**136, 166, 167, 240, **4:**422, 432, 442, 494, **5:**647, **6:**738, 778

excerpts, **2:**241

soil conservation, **3:***365*, **9:***1148–1151*, 1153–1154, *1156*, 1157

agricultural, **1:**8, 10, 13, 106–107

agroforestry, **9:***1176*

animal manure, **1:**33

Bennett's work, **1:**106–107

composting, **2:**210, 211

conservation movement, **2:**213

contamination liability resolution, **1:**29

desertification and, **3:**273

fertilizer and, **4:**486–487

forest operations and, **4:**530

Rodale's work, **8:**1108–1109

sustainable agriculture, **9:**1172–1173

TVA projects, **9:**1202

Spaceship Earth concept, **5:**572–573

Tennessee Valley Authority, **3:**352, **8:**991, **9:**1202–1203, **10:**1315

Thames River restoration, **8:**1086, *1088*

Think Globally, Act Locally, **9:**1212–1213

Thoreau, Henry David, **2:**214, **3:**362, **4:**464, 545, **6:**709, 737, 775, **7:**843, 952, 953, **9:**1214–1215, 1236, 1238, 1239

Three Gorges Project, 5, 685, **5:**682, **9:**1200, 1201, **1216–1219**

Three Mile Island, **2:**205, **6:**705, **7:**895, 897, **9:**1220–1221

Trans-Alaskan Pipeline, **9:**1234–1235

TVA. *See* Tennessee Valley Authority

Ward, Barbara, **6:**778, **10:**1290, **1328–1329**

Wildlife Conservation Society, **3:**373

Wild Life Preservation Society, **3:**373

Wildlife Society, The, **6:**769

Wooing of Earth, The (Dubos), **4:**493, *506*, **10:**1326, 1397

Worldwatch Institute, **6:**780

Law, Government, and Education

Action for Clean Air Committees, **6:**744

Administrative Procedures Act, **7:**897

AFL-CIO Industrial Union Department, **6:**744, 745

Agency for Health Care Policy and Research, **8:**1030

Agency for International Development, **10:**1314

Agency for Toxic Substances and Disease Registry, **8:**1030

Agricultural Commodity and Conservation Service, **10:**1369

Agricultural Conservation Program, **9:**1149

Agriculture Department, U.S., **4:**540, **6:**762, **7:**964, **8:**1021, 1101, **10:**1311, 1313, 1368

regenerative farming research, **8:**1109

soil erosion and conservation programs, **9:**1148, 1149, 1150, 1154, 1157

Air Pollution Control Office, **4:**448

Air Quality Control Act, **2:**179

Alaska National Interest Lands Conservation Act, **7:**852–853, 954

alternative dispute resolution, **1:**26–31

NIMBY issues, **7:**879

Alternative Motors Fuels Act, **5:**557

American Cancer Society, **8:**1030

American Lung Association, **6:**744, **8:**1030

American Museum of Natural History, **6:**796, **7:**849, 967, **8:**1111

Hall of Biodiversity, **1:***113*, **7:***847*, 848

American Ornithologists' Union, **7:**851

American Public Health Association, **10:**1386

American Water Works Association, **10:**1386

Americorps, **10:***1310*

Amnesty International, **10:**1299

Animal Ecology (Elton), **3:**360, 361

Antarctic Treaty (1959), **4:**429

Antarctic Treaty (1992), **1:**46–47, **4:**430

anthropology, **1:**48–53, 63, **7:**872, 943

animism, **1:**43

fire use study, **4:**492

Mead's work, **6:**796–797

tool development study, **9:**1188

Tylor's work, **10:**1286–1287

See also archeology

arbitration, **1:**27

archeology, **1:**62–64

paleontology and, **7:**942

plant domestication finds, **8:**996–997

satellite remote sensing, **8:**1078–1079

Three Gorges Project effects on artifacts, **9:**1219

Argonne National Laboratory, **7:**894

Army Corps of Engineers, U.S., **1:**74–75, **4:**422, **10:**1311, 1313

canal building, **2:**145

cost-benefit analysis, **1:**104

Everglades protection, **4:**469–470

fish ladders, **2:**234–235

functions, **10:**1313

restoration ecology, **8:**1087–1088

water pollution projects, **10:**1369

waterways permits, **4:**422

wetlands preservation, **10:**1372–1373

Associated General Contractors, **7:**903

Association of Schools of Public Health, **8:**1030

Association of Southeast Asian Nations, **4:**427

Atomic Energy Act, **7:**893, 899

Atomic Energy Commission, **4:***423*, **7:**893, 894, 897, **10:**1387

Audubon Society. *See* National Audubon Society

Australian Rangeland Society, **8:**1050

Basel Convention (1992), **4:**429, **10:**1297

BAT standards (best available technology), **9:**1194

Biospheric Aspects of the Hydrological Cycle project, **6:**719

BLM. *See* Bureau of Land Management, U.S.

Bronx Zoo, **3:**373

Brownfields Action Agenda, **2:**139

Brundtland Commission, **4:**453, **9:**1176, 1177, 1178, 1181

Btu tax, **3:**403

Buenos Aires Conference (1998), **10:**1293

Bureau of Biological Survey, U.S., **7:**852, **8:**1021

Bureau of Fisheries, U.S., **8:**1021

Bureau of Labor Statistics, U.S., **7:**899

Bureau of Land Management, U.S., **2:**140–143, **4:**542, **5:**587, **6:**753, 763, **10:**1313

Bureau of Oceans and International Environmental and Scientific Affairs, U.S., **10:**1314

Bureau of Public Roads, U.S., **8:**1101

Bureau of Reclamation, U.S., **5:**587, **6:**729

Cairo Conference (1994), **8:***1014*, **10:***1291*, 1293

Canada. *See* Environment Canada

careers, environment-related. *See* employment and careers

Centers for Disease Control, U.S., **7:**958, **8:**1030

Challenger, HMS, **7:**905, 910

Charleston Museum (S.C.), **7:**846

Children's Fund, UN, **10:***1296*, 1301

CITES. *See* Convention on International Trade in Endangered Species

Citizen's Clearinghouse for Hazardous Waste, **4:**447

Civilian Conservation Corps, **4:**541, **8:**991, **9:**1149

Clean Air Act, **2:**178–179, **9:**1194

acid rain provisions, **1:**4–5

amendments of 1990, **4:**425, **7:**862, **8:**1105, **9:**1232–1233, 1245

application, **4:**420, 421, 448–449

chemical industry, **2:**175

coal-burning compliance, **2:**197–198

commuter vehicle occupancy, **9:**1245

economics of, **3:**338, 339

environmental assessment, **4:**414

EPA implementation, **4:**448–449, 450

history of, **4:**424

industrialization and, **6:**703

labor movement and, **6:**744, 745

natural gas advantages and, **7:**862

pollutant limits, **1:**22–23, 93

pollution trading, **8:**1007

as public interest legislation, **8:**1036

as technology-forcing regulation, **9:**1194

tradable permit programs, **9:**1232–1233

vehicle emissions, **8:**1104–1105

Clean Air Commute (Canadian campaign), **8:**1004, *1005*

Clean Fuel Fleet program, **5:**557

Clean Water Act, **2:**180–181, **10:**1310

applications, **4:**421, **10:**1369

Army Corps of Engineers and, **1:**75

Best Management Practices, **4:**547

environmental assessment, **4:**414

EPA implementation, **4:**450

history of, **4:**424, **10:**1369

provisions, **10:**1369

as public interest legislation, **8:**1036

sport fishing advocates, **4:**497

as technology-forcing regulation, **9:**1194

Climax (Colo.) observing station, **8:**1106

Page numbers in **boldface type** indicate full articles on a subject. Page numbers in *italic type* indicate illustrations or other graphics. All numbers preceding colons indicate volume numbers.

Housing and Urban Development Department, U.S., **7:**891, **10:**1313

Human Environment, UN Conference on the, **10:**1290, 1294, *1295*, 1299

Human Rights Commission, UN, **5:**667

Human Settlements, UN Conference on. *See* Vancouver Conference

IBRD. *See* International Bank for Reconstruction and Development

IGBP. *See* International Geosphere-Biosphere Programme

impact statement, **4:**412, 415, 542–543
    assessment vs., **4:**413
    mining, **7:**838–839
    zoning and land use, **4:**433

Independent Commission on Environmental Education, **4:**418, 434

Inland Waterways Commission, **8:**1111

Intergovernmental Conference on Environmental Education (Tbilisi; 1977), **4:**418

Interior Department, U.S., **4:**541, 542, **5:**587, **7:**891, **8:**1021, **9:**1149, **10:**1313

Intermodal Surface Transportation Efficiency Act, **8:**1102, **9:**1245

Internal Revenue Service, **5:**667

international agreements. *See* environmental law, international

International Atomic Energy Agency, **6:**725, **7:**895

International Bank for Reconstruction and Development, **10:**1294, 1295, 1388

International Biological Program, **10:**1283

International Commission on Radiological Protection, **8:**1045

International Convention for the Regulation of Whaling, **10:**1380–1381

International Convention on the Prevention of Pollution from Ships, **4:**428

International Council of Scientific Unions, **6:**718, **10:**1299

International Court of Justice, **10:**1299–1301

International Dark Sky Association, **6:**773

International Decade for Natural Disaster Reduction, **7:**857

International Development Association, **10:**1388

International Environmental Education Programme, **4:**418

International environmentalism. *See* environmental law, international; United Nations conference; United Nations program

International Federation of Environmental Journalists, **6:**738

International Finance Corporation, **10:**1388

International Geophysical Year, **1:**45

International Geosphere-Biosphere Programme, **6:**718–721

International Global Atmospheric Chemistry project, **6:**718, 721

International Group of Funding Agencies for Global Change Research, **6:**720, 721

International Human Dimensions Programme on Global Environmental Change, **6:**721

International Institute for Environment and Development, **10:**1328

International Joint Commission, **4:**427, 453

International Organization for Standardization, **6:**771

International Register of Potentially Toxic Chemicals, **10:**1297

International Scientific Conference on the Conservation and Utilization of Resources, **6:**769

International Sociological Association, **8:**1039

International Tropical Timber Organization, **4:**431

International Whaling Commission, **10:**1381

International Whaling Convention, **4:**429

IUCN, **6:**733–735
    threatened species list, **3:**370

Joint Global Ocean Flux Study, **6:**718

journalism, **6:**736–739

Justice Department, U.S., **10:**1313–1314

Kyoto Protocol, **4:**453, **10:**1293

Labor Department, U.S., **7:**891, 901, **10:**1314

Laborers' International Union of North America, **7:**903

Lacey Act, **2:**137

Land and Water Conservation Fund, **7:**853

Land Improvement Plan (TVA), **9:**1203

Land-Ocean Interactions in the Coastal Zone project, **6:**719

Land Ordinance of 1785, **6:**751

Land Use and Land Cover Change, **6:**718

law. *See* environmental law *headings*; *specific laws*

Law of the Sea, UN Convention on, **4:**428, **10:**1291–1292

League of Conservation Voters, **4:**443, 447

lobby, environmental, **4:**442–443, 445–446

Long Island Pine Barrens Act, **1:**31

Man and the Biosphere Program, **4:**430, **10:**1296

Manhattan Project, **7:**892

Marine Mammals Act, **4:**501

Marshall Plan, **10:**1328

mediation, **1:**27, 28, 30

meteorology, **6:**816–819
    as biome determinant, **1:**120
    drought and, **3:**300
    oceanography, **7:**911, 913
    remote sensing, **8:**1079
    water cycle, **1:**82–83, **10:**1355, 1357
    weather prediction, **6:**819

microbiology, **7:**956–957

Migratory Bird Conservation Act, **7:**852, **8:**1021

Migratory Bird Hunting Stamp Act, **7:**852

Migratory Bird Treaty Act, **2:**137, **7:**852, **8:**1021

Minerals Management Service, U.S., **5:**587

Mine Safety and Health Act, **7:**902

Mine Safety and Health Administration, **1:**22, **7:**899, 902

Montreal Protocol, **4:**428, **10:**1297

Motor Vehicle Air Control Act, **2:**179

Motor Vehicle Information and Cost Savings Act, **5:**566

moveable museum programs, **7:***848, 849*

Multilateral Investment Guarantee Agency, **10:**1388

Multiple Use-Sustained Yield Act, **4:**540, 542, 543

museums and exhibitions, **7:846–849**

National Academy of Science, **3:**283, **7:**900

National Advisory Environmental Health Committee, **7:**874

National Aeronautics and Space Administration, **5:**586, **7:***938, 939*, **10:**1314

National Ambient Air Quality Standards, **2:**178, 278

National Audubon Society, **1:**89, **4:**423, 436, 442, *445*, 469, **6:**745, **7:**851, 969

National Biological Information Infrastructure, **5:**587

National Center for Atmospheric Research, **8:**1106–1107

National Council on Radiation Protection and Measurements, **8:**1045

National Earthquake Hazards Reduction Program, **5:**586

National Environmental Education Act, **4:**416

National Environmental Education Advocacy Project, **4:**418

National Environmental Policy Act, **4:**433, 450, **5:**667, **6:**753, **8:**1034, 1104, **10:**1311
    air and water pollution requirements, **1:**75, **2:**179, 181
    Clean Fuel Fleet program, **5:**557
    environmental assessment process, **4:**412–415, 542–543
    esthetic values influence, **4:**465
    as first major environmental statute, **4:**424
    highway projects, **8:**1104
    nuclear power plant licensing impact, **7:**896, 897
    technology-forcing regulation, **9:**1194

National Forest Management Act, **4:**465, 540

National Highway System, **8:**1102

National Historic Landmarks, **2:**140

National Institute for Occupational Safety and Health, **7:**874, 900, 901, 903, 904, **8:**1136

National Institutes of Health, **2:**133, **5:**654, **7:**900, **8:**1030

National Interagency Fire Center, **2:**141

National Monuments, **2:**141

National Monuments Act, **8:**1110

National Occupational Exposure Survey, **7:**900

National Oceanic and Atmospheric Administration, **4:**500, **10:**131, 1311

national parks. *See* park, U.S. national

National Parks and Conservation Association, **4:**442, **6:**773

Page numbers in **boldface type** indicate full articles on a subject. Page numbers in *italic type* indicate illustrations or other graphics. All numbers preceding colons indicate volume numbers.

National Park Service, **4:**436, 468, **5:**587, **6:**753, 759, 761, 773, **7:**952–955, **10:**1313
  creation and guiding principle of, **7:**952
  See also park, U.S. national
National Pollutant Discharge Elimination System, **10:**1369
National Register of Historic Places, **2:**140
National Religious Partnership for the Environment, **8:**1066–1067
National Renewable Energy Laboratory, **3:**397
National Resource Conservation Service, **6:**764
National Resources Defense Council, **3:**365, **4:**443, **8:**1034–1035, 1037
National Resources Planning Board, **10:**1387
National Scenic Byways program, **8:**1102
National Science Foundation, **5:**654, **10:**1314
National System of Interstate Highways, **8:**1102
National Toxics Campaign, **4:**447
National Trails System Act, **4:**465
National Wildlife Federation, **4:**442
National Wildlife Refuge Administration Act, **8:**1021
National Wildlife Refuge System, **7:**850–853, *866,* 953, **8:**1021, 1110
National Wildlife Refuge System Improvement Act, **8:**1021
Natural Resources Conservation Service, **1:**107, **6:**762–763, **10:**1311
Natural Resources Defense Council, **4:**423, 437, **8:**1034, 1037
Nature Conservancy, **4:**446, **9:**1212
negotiated rulemaking ("reg-neg"), **1:**27
negotiation, **1:**27
New Deal, **6:**759, **8:**991
  Civilian Conservation Corps, **4:**541, **8:**991, **9:**1149
  conservation movement, **2:**217
  soil conservation, **9:**1147, *1148,* 1149
  suburban housing, **9:**1162, 1163
Newlands Reclamation Act, **2:**181, **8:**1110
New York Fish Commission, **8:**1110
New York Sportsmen's Club, **2:**216
New York State Lung Association, **6:**744
New York Zoological Society, **3:**373, **6:**777
Noise Pollution Control Act, **4:**424, **7:**890
nongovernmental organizations, **2:**247
North American Association for Environmental Education, **4:**418
North American Conservation Conference, **8:**991
North Atlantic Marine Mammal Commission, **10:**1381
Northwest Ordinance of 1787, **6:**751
Nuclear Regulatory Commission, **1:**28, **3:**349, **7:**897, **8:**1045, **9:**1220, **10:**1313
nuisance disputes, **1:**30–31
nuisance law, **4:**433

occupational safety and health, **5:**631, 633–636, **7:**898–903
  air pollutants, **1:**19, 22
  asbestos risks, **1:**77
  carbon monoxide poisoning, **2:**161
  chemical industry, **2:**172, 174
  factory safety laws, **7:**898–899

Nelson's work, **7:**874
  petroleum industry, **8:**982–983
  radon risk, **8:**1044, 1046
  sick building syndrome, **9:**1136–1137
  workers' compensation, **2:**179
Occupational Safety and Health Act, **6:**744, 745, **7:**899, 901
Occupational Safety and Health Administration, **1:**22, 77, **2:**161, 174, **7:**703, 900, 901–904, **10:**1314
  Environmental Network, **6:**745
  establishment of, **7:**899
Ocean Dumping Convention, **4:**427, 428
Oceanic Society, **4:**443
oceanography, **7:**904–913
  Carson's contribution, **2:**166
  Cousteau's contribution, **2:**231
  geologic mapping, **5:**585, 586
  international projects, **6:**718
  sulfur cycle, **6:**833
Office of Economic Ornithology, **8:**1021
Office of Environmental Education, **4:**416, 417
Office of Management and Budget, **10:**1311
Office of Science and Technology Policy, **10:**1311
Office of Surface Mining, U.S., **5:**587
Office of the U.S. Trade Representative, **10:**1311
Oil Pollution Act, **6:**736
1000 Friends of Oregon, **8:**1035
OPEC, **1:**69, **3:**352, **7:**918–923
Organization for Economic Cooperation and Development, **8:**1038, 1039
Organization of Petroleum Exporting Countries. *See* OPEC
Origin of the Alps, The (Suess), **9:**1165
Ornithological Biography (Audubon), **6:**775
Our Common Future (1987 report), **2:**217

paleontology, **7:**942–943
  evolution and, **4:**477, **7:**943
  geology and, **5:**582, 583
park, U.S. national, **2:***212, 213,* **6:**753, 759, 761, 773, **7:**917, **952–955**
  conservation movement and, **2:**214, 216
  esthetics, **4:**465
  Everglades, **4:**468–469, 470
  old-growth forests, **4:**541
  Roosevelt's (Theodore) establishment of, **8:**1110
  smog in, **9:**1142
  See also forest, U.S. national; National Park Service
Past Global Changes Study, **6:**719, 721
Pattern of Animal Communities (Elton), **3:**360
petrology, **5:**585
physiography, **6:**748–749
Pinchot Institute for Conservation Studies, **8:**991
Pollution Probe, **1:***19,* **8:**1004–1005
Population and Development, UN Conference on, **8:***1014,* **10:***1291,* 1293
Presidential Council on Environmental Quality, **4:**424
press. *See* journalism
Principles of Geology (Lyell), **5:**583, **6:**784, 785
Protection Act (1894), **8:**1110

Public Health Service, U.S., **2:**178, **7:**874, **8:**1030, **10:**1276, 1369
  agencies of, **8:**1030
  National Institute for Occupational Safety and Health, **7:**901
  nitrate levels, **7:**882
  noise exposure risks, **7:**890
public interest, **8:1034–1037**
  acid rain problem, **1:**4–5
  environmental law and, **4:**424–425
  environmental literacy as, **4:**434–435
  national forests and, **4:**540–543
  pricing system and, **8:**1024–1025
  risk, **8:**1099–1095
public lands, **6:**752–753, 763, 765, 776
  Bureau of Land Management, **2:**140–141
  conservation movement, **2:**213–214, 216
  historical misuses of, **2:**213
  See also forest, U.S. national; National Wildlife Refuge System; park, U.S. national
Public Utility Holding Company Act, **3:**352
Public Utility Regulatory Policies Act, **3:**350, 352, 401

Raker Bill (1913), **8:**1018
Ramsar Convention (1975), **4:**430
Reclamation Act, **6:**729
Red Data Books, **3:**370
Regional Clean Air Incentives Market, **9:**1232
Regional Research Networks, **6:**721
René Dubos Center for Human Environments, **4:**434, **9:**1212
Resource and Environmental Profile Analysis, **6:**770
Resource Conservation and Recovery Act, **2:**138, **4:**424, 450, **5:**626, **9:**1194, **10:**1342
Responsible Care Program, **1:**109
Rio Declaration (1992), **2:**217, **4:**431, **5:**667, **10:**1290, 1292–1293
  and healthy environment as human rights issue, **5:**667
  See also Earth Summit
Rio Summit (1992). *See* Earth Summit
riparian rights, **4:**433
Rivers and Harbors Act, **2:**181, **4:**422
Rodale Press, **8:**1108–1109
Roosevelt, Franklin D., **8:**990, 991, **9:**1162
  conservation movement and, **2:**216–217
  Manhattan Project, **7:**892
  soil conservation program, **9:**1149, 1150, 1157
  See also New Deal
Ruckelshaus, William, **4:**448, 449, 451

Safe Drinking Water Act, **2:**181, **4:**424, 450
satellites
  Earth Observing System, **3:**308–311, **6:**709, **8:**1079
  El Niño tracking, **3:***356, 357, 358–359*
  remote sensing, **8:**1076–1079
  weather, **6:**819, 820
Scenic Hudson Preservation Conference, **4:**422
Scientific Certification Systems, **3:**314, 315
Scientific Conference on the Conservation and Utilization of Resources, UN, **8:**991

Sea Shepherd Society, **4:**440, 446
Secretariat, UN, **10:***1300*, 1301
Security Council, UN, **10:**1299, *1300, 1301*
Senate, U.S., committees, **4:**424, **10:**1312, 1315
Sierra Club, **2:**214, **4:**436, 437, 442, *447*, 461, 530, **5:**619, **6:**744, 745, **7:**843, **8:**1018
    Legal Defense Fund, **4:**423, 447, **8:**1034, 1036, 1037
SINET (newsletter), **8:**1039
Smithsonian Institution, **6:**790, **10:**1314–1315
Social Indicators Development Program (OECD), **8:**1039
Social Indicators Research (journal), **8:**1039
Social Security Act, **7:**899
Society for Range Management, 1050, **8:**1050
Society of Environmental Journalists, **6:**738
sociobiology, **7:**873
Soil and Water Conservation Society, **1:**107
Soil Conservation Service, U.S., **1:**106–107, **4:**541, **9:**1149, **10:**1369
Soil Erosion Service, U.S., **9:**1149
Solid Waste Disposal Act, **10:**1335, 1341
South Coast Air Quality Management Plans, **8:**1105
Southern California Air Quality Management District, **8:**1007
Southern Environmental Law Center, **8:**1035
Species Survival Commission, **6:**733
Standard State Soil Conservation Districts Law, **9:**1150
stargazing, **6:**773
START Regional Research Networks, **6:**720–721
State Department, U.S., **10:**1314
State Environmental Policy Acts, **4:**412
Statue of Liberty, **7:**955
Stockholm Conference (1972), **3:**304, 305, 307, **4:**414, 418, 426, 427, 431, **8:**1076, **9:**1212, **10:**1290, 1294, *1295*, 1328
Substance Abuse and Mental Health Services Administration, **8:**1030
Superfund, **1:**29, **4:**424, 444, *451*, **9:***1168–1169*
    brownfields cleanup, **2:**138
    EPA implementation, **4:**450, 452, **9:**1168–1169
    hazardous industrial wastes, **5:**628–629
    labor movement and, **6:**745
    Love Canal cleanup, **6:**783, **9:**1168
Supreme Court, U.S., **8:**1037, **10:**1315, 1399, 1400, 1401
Surface Mining Control and Reclamation Act, **2:**197
Surface Transportation Assistance Act, **8:**1102
Sustainable Business Challenge exam, **4:**435
System for Analysis, Research, and Training, **6:**720–721

Tableau Economique, **8:**1023
Task Force on Global Analysis, Interpretation, and Modeling, **6:**719–720
taxes. *See* energy tax; green taxes

taxonomy, **1:**96
Tbilisi Declaration (1977), **4:**418
technology-forcing regulation, **3:**339, **9:1194–1195**
    economic incentive, **3:**332–333
    energy tax, **3:**402–403
    pollution trading, **4:**425
Tennessee Valley Authority, **3:**352, **4:**541, **8:**991, **9:1202–1203**, **10:**1315
TERRA (remote sensing spacecraft), **3:**308–310, 311, **8:**1079
Texas Railroad Commission, **7:**919
Theory of Moral Sentiments, The (Smith), **9:**1138–1139
Theory of the Earth (Lyell), **6:**784
Think Globally, Act Locally, **9:1212–1213**
Three Gorges Project, **9:1216–1219**
Toxics Release Inventory, **9:**1169
Toxic Substances Control Act, **2:**147, **4:**424, 450, **6:**744, 745
Train, Russell, **4:**449
Transportation Department, U.S., **4:**422, **5:**566, **8:**1101, **10:**1314
Transportation Equity Act for the 21st Century, **8:**1102
Trusteeship Council, UN, **10:**1299, *1300*
TVA. *See* Tennessee Valley Authority

UNEP. *See* Environment Programme, UN
UNESCO. *See* Educational, Scientific, and Cultural Organization, UN
United Nations conference, **4:**431, **9:**1212, **10:1290–1293**, *1295*, 1298, 1328
    climate change, **4:**428
    conservation, **8:**991
    desertification, **3:**270, **4:**431
    environmental education, **4:**416
    environment and development. *See* Earth Summit
    global climate change, 1293
    human environment, **3:**304, 305, 307, **4:**414, 418, 426, 427, **5:**667, **8:**1076, **9:**1212, **10:**1290, 1294, *1295*, 1328
    human settlements, **10:**1290–1291
    population and development, **8:***1014*, **10:***1291*, 1293
United Nations Family Planning Agency, **8:**1014
United Nations programs, **10:1294–1297**
    environment, **4:**434, **8:**1066, **9:**1176, 1212
    environmental assessment, **4:**414
    environmental protection, **4:**426–431
    environment and development, **2:**247, **10:**1294–1295, 1301, 1388–1389
    family planning, **8:**1014
    forest resources, **2:**254, 255
    health care, **8:***1040*
    Human Rights Commission, **5:**667
    Lake Baikal, **1:**100, 101
    list of, **10:**1297
    natural disaster, **7:**857
    refugees, **8:***1029*
    social indicators, **8:**1038
United Nations system, **10:1298–1301**, 1388, 1389

University of Chicago, **7:**893, 894
University of Colorado, Boulder, **8:**1106
University of Wisconsin-Stevens Point, **4:**428
U.S. government, **10:1310–1315**
    air pollution regulation, **1:**22–23, **2:**178–179, 197–198
    alternative dispute resolution, **1:**28–29
    alternative fuel promotion, **5:**555, 557
    Army Corps of Engineers, **1:**74–75
    bird protection, **2:**137
    brownfield cleanup, **2:**138, 139
    carcinogen regulations, **2:**147
    clean coal program, **2:**198, 199
    conservation policies, **2:**216
    document interpretation, **2:**178
    Earth Day inception, **3:**306
    ecolabeling standards, **3:**315
    electric utility regulation, **3:**349, 352
    endangered species protection, **3:**372, 374–375
    energy efficiency policy, **3:**401
    energy tax, **3:**402–403
    environmental assessment, **4:**412–415
    environmental careers, **3:**369
    environmental education, **4:**416–419
    environmental law bases, **4:**432–433
    environmental policies, **3:**337–339, 350
    Everglades restoration, **4:**468–470
    fish and wildlife conservation record, **8:**1021
    Fish and Wildlife Service, **4:**494–495
    Geological Survey, **5:**583, 586–587
    highway system, **8:**1101–1102
    Human Genome Project, **2:**133
    irrigation projects, **6:**729
    land ownership, **6:**752–753, 763, 765
    land use, **2:**140–141, 213, 214
    land use controls, **6:**763–764
    national forests, **4:**540–543
    national park system, **7:**917, 952–955
    natural resources management chronology, **8:**1021
    NIMBY policies, **7:**878
    nuclear power, **7:**892–893, 896–897
    occupational safety and health measures, **2:**174, 175, **7:**874, 899, 901–902, 903
    public health practice, **8:**1031–1032
    soil conservation programs, **1:**106–107, **9:**1147, *1148*, 1157
    Strategic Petroleum Reserve, **8:**981
    technology-forcing regulation, **9:**1194–1195
    Tennessee Valley Authority, **9:**1202–1203
    waste management, **10:**1341–1342, 1343
    water pollution control, **10:**1368–1369
    water quality regulation, **2:**180–181
    wildlife refuge system, **7:**850–853
    See also environmental law *headings*; Environmental Protection Agency; *other specific departments and agencies*

Vancouver Conference (1976), **10:**1290–1291
Vienna Convention for the Protection of the Ozone Layer (1985), **4:**427–428

Walsh-Healey Public Contracts Acts, **7:**899, 901

Page numbers in **boldface type** indicate full articles on a subject. Page numbers in *italic type* indicate illustrations or other graphics. All numbers preceding colons indicate volume numbers.

Water Erosion Prediction Project, **9:**_1150_
Water Pollution Control Administration, **4:**448
Water Pollution Control Amendments. _See_ Clean Water Act
Water Quality Act, **2:**161, 181, **4:**450
Water Resources Act, **1:**104
Watt, James, **4:**444, 449, **7:**954, **9:**1191
White House Conference on Natural Beauty, **4:**465
Wild and Scenic Rivers Act, **4:**465
Wilderness Act, **4:**465
Wilderness Society, **6:**769
Wild Horse and Burrow Act, **2:**142–143
World Bank, **9:**1177, 1217, **10:**1294, 1295, 1296, 1299, 1314, 1328, **1388–1389**
World Business Council for Sustainable Development, **4:**435
World Climate Research Programme, **6:**721
World Commission on Environment and Development, **2:**217
World Commission on Protected Areas, **6:**733
World Conference on Education for All, **10:**1296
World Conservation Conference, **8:**991
World Conservation Congress, **6:**734
World Conservation Union. _See_ IUCN
World Directory of Environmental Organizations, **3:**365
World Fertility Survey, **8:**1011
World Health Organization, **6:**717, 725, **7:**874, **8:**1028, 1030, _1040_, **10:**1294, 1295, 1298–1299, 1387
definition of word _health_, **8:**1028
dioxins, **3:**283
food additives, **3:**278
irradiated food safety, **6:**725
tuberculosis control, **10:**1277
World Heritage Fund, **4:**430, **10:**1296
World Heritage List, **10:**1296
World Hunger Project, **2:**165
World Meteorological Organization, **10:**1294, 1295
World Resources Institute, **9:**1155
World Trade Organization, **6:**_744_, 745
Worldwatch Institute, **9:**1151
Worldwide Light Pollution Study, **6:**773
World Wide Web, museum programs, **7:**849
World Wildlife Fund, **2:**247

Yale School of Forestry and Environmental Studies, **8:**991, 1111

zoning, **4:**433, 505–506, **10:1398–1401**
housing discrimination, **5:**645
land ownership and, **6:**754
suburban, **10:**1309, 1399, 1401
urban, **10:**1305, 1307–1308, 1398–1401

Art, Literature, and Ethics

abolitionism, **9:**1238
aesthetics. _See_ esthetics
Alcott, A. Bronson, **9:**1236, 1238–1239
animism, **1:42–43**
anthropocentrism, **4:**437–438, 460, 462
archetypes, **7:**873, 878

architecture, **1:65–69**
art esthetics, **4:**463–464
Audubon bird paintings, **1:**88–89, **2:**214, _215_
landscape painting, **4:**463–464
Audubon, John James, **1:**88–89, **2:**214, _215_, **6:**709, 775
Augustine, Saint, **8:**1074
Austin, Mary, **6:**777

Benedictines, **8:**1074, 1075
Birds of America, The (Audubon), **2:**214, _215_
Birds of North America, The (Audubon), **1:**88–89, **6:**775
Black Elk Speaks (Black Elk), **1:**43
Blake, William, **8:**_1075_
Borland, Hal, **6:**779
Breuer, Marcel, **1:**68
Buddhism, **5:**648, **8:**1062, 1063–1064, 1065, 1068–1069, _1071_

cave painting, **6:**708, _709_
Celebrations of Life (Dubos), **9:**1212
Celtic Christianity, **8:**1074, 1075
Christianity, **8:**1065, 1067, 1068, 1073–1075
Civil Disobedience (Thoreau), **9:**1215
Cole, Thomas, **2:**214, **4:**464
Coleridge, Samuel Taylor, **9:**1239
Compleat Angler, The (Walton), **4:**469
Concord Summer School of Philosophy, **9:**1238–1239
Cooper, James Fenimore, **2:**214, **4:**464, **6:**_774_, 775
creation beliefs, **8:**1065, _1067_, 1073–1074, 1075

deep ecology, **2:**215, **248–249**, **4:**436
Doxiades, Constantine, **3:296–97**
Druidism, **8:**1074
dystopianism, **10:**1317

ecofeminism, **3:312**
ecotourism, **9:**1249, 1252–1253
Ectopia (Callenbach), **10:**1317
Eightfold Path, **8:**1068
Emerson, Ralph Waldo, **2:**214, 219, **3:362–363**, **6:**709, 775, **7:**843, 952, **9:**1214, 1236, _1237_, 1238, 1239
Enlightenment, **2:**212, **6:**788, **9:**1236
equity, **4:460–462**
animal rights as, **1:**38–41
environmental assessment and, **4:**415
environmental risk and, **4:**438, 460–462
housing discrimination, **5:**645
intergenerational, **2:**217, **3:**339, **9:**1176
land ownership, **6:**755
NIMBY issue, **7:**878
urban park, **7:**948, 949
Eskimo, **5:**594, **10:**1282, 1283
esthetics, **4:463–465**
biodiversity and, **1:**113
humanized environment, **5:**656–661
landscape, **6:**756–761
as one of five e's, **4:**_502–503_, 504–506
outdoor recreation and, **8:**1056–1057, 1059
urban forests, **4:**536
zoning for, **10:**1400

ethics
cloning, **2:**193
genetic engineering, **3:**293
land use, **6:**753, 769, 777–778
as one of five e's, **4:**506–507
See also religion

Face of the Earth, The (Suess), **9:**1165
feng shui, **5:**588–589, **6:**_749_
Field Guide to the Birds, A (Peterson), **7:**969
five e's, **4:502–507**
Fuller, Margaret, **9:**1236, 1238

Gaia (deity), **1:**42
Genesis, **8:**_1072_, 1073, 1074, 1075
Golden Fish, The (Cousteau film), **2:**231
Greek philosophy, **3:**354
green belt, **5:608–609**, **9:**1163, **10:**1308
Ground of the City of Vienna, The (Suess), **9:**1164

heredity vs. environment. _See_ nature vs. nurture controversy
Herodotus, **5:638–39**
Hinduism, **5:**648, **8:**1063–1064, 1065, 1069–1070
historic districts, **4:**464–465
human development. _See_ nature vs. nurture controversy
human ecology, **5:648–651**
anthropology and, **1:**49, 50–53
architecture and, **1:**66–69
Arctic River basin, **1:**72–73
biodiversity and, **1:**114
biosphere, **2:**128–129
ecosystem changes, **3:**344–346, 347
environmental movement and, **4:**437–438
equity and, **4:**460–461
five e's and, **4:**502–507
human rights and, **5:**664–667
indigenous peoples and, **5:**690–691
population changes, **3:**_322–323_
human rights, **5:664–667**

indigenous peoples, **5:688–691**
coniferous forests uses, **4:**524
fire uses, **4:**492–493, 528, 545
fuelwood, **5:**_568_, 569
Great Barrier Reef, **5:**504
land ownership views, **5:**689, **6:**750–751
medicine, **6:**800–801, 805
rangeland use, **8:**1050, _1052_, 1053
religion, **8:**1062–1063, 1065, 1074–1075
tundra, **10:**1282–1283
See also Native Americans
intergenerational equity, **2:**217, **3:**339, **9:**1176
Inuit, **1:**53, **8:**1062, **10:**1282, 1283
Islam, **7:**923, **8:**1065, 1068, 1070–1071, 1072
Jainism, **8:**1063–1064, _1066_
Judaism, **8:**1065, 1067, 1068, 1072–1073, 1075

karma, **8:**1068
land ownership, **6:754–755**
green belt, **5:**609
indigenous peoples, **5:**689, **6:**750–751

Page numbers in **boldface type** indicate full articles on a subject. Page numbers in *italic type* indicate illustrations or other graphics. All numbers preceding colons indicate volume numbers.

vegetarianism, **1:**40
visual art. *See* art esthetics

Walden (Thoreau), **2:**214, **3:**362, **6:**737, 775, **7:**953, **9:**1214–1215
Wealth of Nations (Smith), **8:**1023, **9:**1138, 1139
well-being. *See* quality of life
Whitman, Walt, **9:**1236, *1237*
Whittier, John Greenleaf, **9:**1238
wilderness experience, **6:**768–769, **7:**843, **10:**1382–1383
  ecotourism, **9:**1249, 1252
  preservationism and, **8:**1016, 1036
  recreation and, **8:**1056
  Walden Pond, **9:**1214–1215
  wilderness areas, **4:**541, 542–543, 546
Wilderness Society, The, **4:**443, **6:**769, **7:**843
World without Sun (Cousteau film), **2:**231
Wright, Frank Lloyd, **1:**68, **10:**1390–91

yin and yang, **5:**648
yoga, **6:**805–806, 807

Zen Buddhism, **6:**757, 781, **8:**1068
ziggurat, **10:**1396–1397
Zoser pyramid, **1:**65

Index of Places

Adirondack Park, **6:**775
Adriatic Sea, **2:**243, **4:**510
Afghanistan
  animal domestication, **1:**33
  irrigation canal, **1:**13
  refugees, **6:**826
Africa
  AIDS, **1:**17, **8:**1013
  contraceptive types, **4:**484
  Ebola virus, **4:**458
  hearths, **4:**490
  high birth and death rates, **8:**1009, 1011, 1012
  Masai cattle herds, **8:**1052
  refugees, **6:**825
  seawater farms, **7:**924
  tropical deforestation, **2:**254
  UNICEF health center, **8:**1040
  See also North Africa; *specific countries*
Alabama, Muscle Shoals, **9:**1202, 1203
Alameda Park (Mexico City), **5:**661
Alaska, **4:**434
  Aleutian Islands, **6:**740, **10:**1379
  Arctic National Wildlife Refuge, **6:**736
  Ballinger-Pinchot controversy, **8:**980
  coniferous forests, **4:**524
  conservation, **7:**852–853, 954, **8:**990
  EPA water pollution check, **4:**449
  fishing industry, **4:**498
  Green Party, **5:**619
  indigenous peoples, **10:**1282–1283
  national parks, **7:**954
  oil exploration, **9:**1234, 1235, **10:**1278, 1283
  oil spills, **6:**705, 736, **7:**914, **9:**1235
  permafrost, **10:**1280
  pipeline, **9:**1234–1235, **10:**1283
  shoreline riprap, **9:**1132

temperate rain forest, **9:**1197
Tongass National Forest, **4:**540
tundra, **5:**594, **10:**1278, *1279*, 1282
USGS data collection, **5:**586
wildlife refuges, **7:**6.736, 850, 851, 852–853, **8:**1036
Alaskan Maritime NWR, **7:**851
Albany (N.Y.), **7:**917
Aleutian Islands, **6:**740, **10:**1379
Alexandria (Egypt), **9:**1134
Alexandria (Va.), **8:**1100
Algeria, **2:**263, **7:**920, 921, 923
Allalin Glacier, **5:**595
Alps, **9:**1251
Amazon basin, **1:**113, **10:**1272, 1273, 1274
  ants and termites, **9:**1144
  deforestation, **2:**255
  medicinal plant, **6:**799
  rubber, **8:**999
  Wallace naturalist expedition, **10:**1322
Amazon River, **10:***1275*
  floating houses, **1:**6
  tropical rain forest, **10:**1264
American River, **1:**75
Amherst (Mass.), **10:**1376
Anaheim (Calif.), **6:**726
Anchorage (Alaska), **5:**586, **6:**736
Andes ice fields, **5:**580
Angara River, **1:**100
Angola, **6:**825, 826
Anhui Province (China), **10:**1325
Annapolis Royal (Nova Scotia), **3:**395
Antarctica, **1:**44–47, 85
  bamboo
  Earth Day activities, **3:**306
  international environmental treaty, **1:**46–47, **4:**429–430
  mining protocol, **5:**597
  oceans, **7:**912
  ozone hole, **2:**169, **4:**428, **7:**940
  tundra, **10:**1278
  whaling, **4:**430, **10:**1380, *1380*, 1381
Appalachia, **2:**204, 213
Appian Way, **8:**1100
Aral Sea, **1:**60–61, **6:**705
Arches National Park, **7:**955
Arctic Circle, **3:**360, **9:**1234, **10:***1289*
Arctic National Wildlife Refuge, **6:**736, **9:**1235
Arctic Ocean, **9:**1132
Arctic River basin, **1:**70–73, *73*, **10:**1278
Arctic Village (Alaska), **10:**1283
Area de ConservaciÛn Guanacaste (Costa Rica), **8:**1086, 1087, *1089*, **9:***1256*
Argentina
  agricultural technology, **9:**1192
  birthrate decline, **8:**1010
  cattle raising, **5:**600
  pampas, **1:**119, **5:**600, **8:***1049*
  ranches, **8:**1048, *1049*
Arizona
  Biosphere 2, **3:**345
  canal irrigation, **2:**144, **6:**729
  Casa Grande, **1:**42
  Earth Day activities, **3:**307
  electric vehicle fueling station, **5:**556
  Grand Canyon, **5:**586, **9:**1127, 1143, 1201
  mineral production, **7:**841

pest inspection and quarantine, **7:**964
rangeland, **8:**1049, 1053
tundra, **10:**1278
Wright's Taliesin West, **10:**1391
Arkansas, wildlife refuge, **6:**696
Arkansas River, **1:**75
Arlington (Va.), **10:**1304–1305
Armenia, earthquake, **7:**857
Armley Mill Industrial Museum (Leeds, England), **6:**702
Arnold Arboretum (Boston), **7:**917
Arno River, **4:**510
Arroyo Seco Freeway, **8:**1103
Aruba, desalination plant, **2:**259
Asheville (N.C.), **7:**917
Aswan High Dam, **1:**80–81, **5:**639, 682
Athens (Greece), **3:**297, *297*
  ancient waste disposal, **10:**1330
  Temple of Athena Nike, **1:**65–66, *65*
Atlanta (Ga.), **9:**1242, 1246
Atlantic Ocean, **10:**1358
  tidal power, **3:**395
  U.S. fishing industry, **4:**500, 501
Austin (Tex.), 1385
Australia
  Aborigines, **8:**1062, *1063*
  agricultural technology, **9:**1192
  chaparral, **2:**171
  El Niño effects, **3:**357
  GDP per capita, **8:**1027
  grasslands, **1:**123, **5:**598
  Great Barrier Reef, **5:**604–605, **9:**1251–1252
  Hayman Island Resort, **9:**1251–1252
  ranches, **8:**1048
  rangeland, **8:**1050, 1053, 1054
  temperate rain forest, **9:**1196
  tropical dry forest, **9:**1256
  unique biota, **2:**130, *131*
Austria, **5:**619, **9:**1246

Babylon, **6:**757, **8:**1100, **10:**1397
Bahamas, **9:***1130*
Baikal, Lake, **1:**100–101, **9:**1129
Bali, rice paddies, **1:**10
Baltic Sea, **2:**243
Baltimore (Md.)
  commuter Clean Air requirements, **9:**1245
  rapid transit system, **9:**1242
  waterfront development, **4:**414, **10:**1303, *1361*
Bangladesh
  bamboo, **1:**103
  family planning program, **8:**1010
  floods, **4:**510, **7:**854, 856
  Grameen Bank, **5:**671
  hurricanes, **7:**913
  percent of underweight children, **5:**673
Baraboo (Wis.), **6:**768
Barbados, **9:**1157
Barcelona (Spain), urban space project, **7:**950–951
Battery Park City (N.Y.C.), **10:**1302, 1361
Bay of Fundy, **3:**395, **10:**1352
Beijing (China), **1:**115
  Tiananmen Square, **5:**664–665
Belen (Brazil), **10:***1275*

Belgium
    dioxin in food chain, **3:**283
    Green Party, **5:**619
    North Sea flood, **4:**510
    shoreline settlement, **9:**1134
Belle Ayr open-pit mine (Wyo.), **7:**840
Beni Biosphere Reserve (Bolivia), **2:**247
Benson Creek (Wash.), **10:***1371*
Bering Land Bridge, **5:**594–595
Bering Sea, **4:**498, **5:**594, **10:**1379
Bering Strait, **10:**1282
Bhopal, **1:108–109**, **2:**205
Big Cypress Preserve, **10:***1362*
Big Cypress swamp, **4:**469, 470
Big Springs (Neb.), **8:***1101*
Birkenhead Park (Liverpool, England),
    7:948
Biscayne Bay, **4:**469, 471
Black Forest, **9:**1251
Black Sea, **2:**236, 243, **7:**936
Blue Ridge Mountains, **8:**1100
Bolivia, **3:***404*
    Beni Biosphere Reserve, **2:**247
    fertility rate, **8:**1010
Boston (Mass.), **10:**1309
    Charles River Natural Valley Storage
        Project, **10:**1372–1373
    green belt circuit, **5:**609, **7:**949
    Inner Traffic Loop, **8:**1104
    as Northeast corridor end, **9:**1243
    parks and community gardens, **4:***549*,
        5:609, **7:**917, 949
    rapid transit system, **9:**1242
    waterfront development, **10:**1303, 1361
    waterwheel, **5:***682*
Botswana
    demographic transition, **8:**1009
    Kalahari Desert, **6:***824*
Boulder (Colo.), **5:**609, *616*
Boundary Waters Canoe Area (Minn.),
    3:321–323
Bowling Green (Ky.), **5:***622*
Brazil, **2:***203*, 205
    birthrate decline, **8:**1010
    deforestation, **2:**2.*253*
    GDP in billions of dollars, **8:**1027
    green revolution, **5:**620
    per capital energy consumption, **9:**1179
    percent of underweight children, **5:**673
    planned community, **10:**1302
    ranches, **8:**1048
    Rio de Janeiro, **2:***203*, **5:**561, **6:***810*
    Roosevelt (Theodore) expedition, **8:**1111
    rubber, **8:**999
    scrub savanna, **1:**119
    tin mine, **2:***255*
    tropical dry forest, **9:**1256
    tropical rain forest, **10:**1264
    See also Amazon basin; Amazon River
Bristol Bay, **7:**851
Britain. *See* Great Britain
British Columbia, **4:**453, **6:***749*, **9:***1167*
Brittany (France), **3:**395
Bronx River Parkway, **8:**1103
Brook Farm, **3:**362

Brooklyn (N.Y.), **7:**917, 949, **9:***1135*
Broward County (Fla.), **5:***629*
Brunswick (Ga.), **10:***1346*
Bryce Canyon (Utah), **7:***953*
Buffalo Creek Disaster, **2:**204
Burkina Faso, **5:***645*

California
    annual grasslands, **8:**1053
    chaparral, **1:**19, **2:**171, **8:**1053
    chemical industry storage, **1:**109
    desalination plant, **2:**258
    earthquake damage, **5:***584*, **7:**856, 857
    El Niño effects, **3:**357, **7:**912
    emissions controls, **1:**92, 93, **6:**703, **7:**862,
        9:1126, *1142*, 1232, **10:**1392, 1393
    energy efficiency, **3:**399
    environmental assessment law, **4:**3
    exotic plants, **4:**481
    geothermal energy, **3:**391
    gold rush, **5:**601
    Green Party, **5:**619
    Hetch Hetchy damming debate,
        8:990–991, 1016, 1018, *1019*
    highways, **5:***562*
    household toxic waste, **5:**625
    irradiated fruit marketing, **6:**726
    irrigation, **2:**144, **6:**731, 732
    Kings Canyon, **4:***447*
    Klamath weed control, **6:**716
    Lassen Volcanic National Park, **6:***832*
    literature on, **6:**776, 777
    migrant workers, **6:**823
    mineral production, **7:**841
    Muir Woods National Monument, **8:**991,
        1110
    natural disasters, **7:**856
    organic farming, 1174
    Pacific prairie, **1:**122
    pest inspection and quarantine, **7:**964
    planned community, **10:**1302
    pollution trading, **8:**1007
    ranching, **8:**1048
    rangeland, **8:**1049, 1053
    Redwood National Park, **4:***438*, **7:***954*
    Sacramento levee construction, **1:***75*
    San Diego sewage processing, **10:**1363
    San Joaquin Valley, **9:**1124–1125
    Santa Barbara oil spill, **4:**423
    sardine fishery collapse, **7:**907
    Sequoia–Kings Canyon National Park,
        8:991
    Sierra Nevada, **7:**842, 843, **9:**1124, *1125*,
        10:1278
    smog, **9:***1141*
    soil salinization, **6:**731
    suburban development, **9:***1161*
    tradable permit, **9:**1232
    waste disposal, **10:**1333
    wind turbines, **3:**397
    zero-emission vehicles, **10:**1392, 1393
    See also Los Angeles; San Francisco;
        Yosemite National Park
Cambodia, **10:***1326*
Camden (N.J.), **10:**1309

Canada
    acid rain, **1:**3, 4
    agricultural technology, **9:**1192
    coniferous forests, **4:**524
    Earth Day activities, **3:**307
    ecolabeling, **3:**314
    environmental law, **4:**453–454
    Environment Canada, **4:**453–454
    GDP per capita, **8:**1027
    Great Lakes, **5:**606
    Lyell Icefield, **7:***936*
    natural gas reserves, **7:**858
    Ontario suburban housing, **5:***644*
    Pollution Probe, **1:***19*, **8:**1004–1005
    Prince Edward Island solar house, **3:***400*
    ranching, **8:**1048
    Responsible Care Program, **1:**109
    TERRA joint project, **3:**308, **8:**1079
    tidal power, **3:**395
    totem pole, **1:***49*
    tundra, **10:**1278, 1279, 1283
    waste disposal, **10:**1336
    wildflowers, **10:**1385
    See also specific provinces
Cape Fear (N.C.), **8:**1035
Caribbean region, **9:***1130*
    dengue hemorrhagic fever, **3:**287
    desalination plants, **2:**258
    hurricanes, **7:**913
Cartago (Costa Rica), **3:***350*
Casa Grande (Arizona), **1:***42*
Cascade Mountains, **7:**950, **8:**1053, **9:***1170*
Cascades National Park, **7:***952*
Cascadian Farm (Wash.), **9:**1170
Caspian Sea, **2:**145
Çatal Hüyük (Turkey), **1:***34*
Central Arabian Arch aquifers, **6:***730*
Central Park (N.Y.C.), **4:**464, **6:**758, 761,
    7:916, *917*, 949, 953
Chad, **8:**1011
Ch'ang River (Yangtze River), **9:**1199–1200,
    1201, 1216, *1217*, 1218, 1219, **10:**1324
Chantilly (France), **6:**757
Charles River Natural Valley Storage Project,
    10:1372–1373
Charleston (S.C.), **4:**464, **7:**846
Chattanooga (Tenn.), **9:**1246
Chengdu (China), **9:**1200
Chequamegon National Forest, **8:***1058*
Chernobyl, **1:**73, **2:176–77**, 205, **6:**705,
    7:896, **8:**1046, **10:**1326
Chesapeake and Ohio Canal, **1:**74
Chesapeake Bay, **2:**242, **3:**275, **6:**830,
    10:1368
Chiapas (Mexico), **5:**666
Chicago (Ill.)
    commuter Clean Air requirements,
        9:1245
    garden suburb, **7:**917, **10:**1307
    park system, **7:**949
    rapid transit system, **9:**1242
    slaughterhouses, **8:**1048
Chicago World's Fair, **7:**917
Chihuahuan Desert, **8:**1053
Chile
    birthrate decline, **8:**1010
    fishing industry, **4:**498
    IUCN, **6:**733
    temperate rain forest, **9:**1196

Page numbers in **boldface type** indicate full articles on a subject. Page numbers in *italic type* indicate illustrations or other graphics. All numbers preceding colons indicate volume numbers.

China
    alternative medicine, **6:**802, 804, 805
    ancient waste disposal, **10:**1331
    aquaculture, **1:**56
    bamboo, **1:**103
    biogas digesters, **1:**115
    Bronze Age weapons, **9:***1188*
    Buddhism, **8:**1068
    chrysanthemum reverence, **8:**999
    Communist planned community,
        10:1317
    demographic transition, **8:**1009
    dengue hemorrhagic fever, **3:**286
    domesticated animals, **1:**35–36
    double cropping, **9:**1171
    Earth Day activities, **3:**307
    earthquake, **7:**856
    erosion rates, **9:**1153, 1156–1157
    family planning program, **4:**483, 484, 485,
        8:*1010*
    feng shui (geomancy), **5:**588–589, **6:***749*
    first plant domestication, **8:**996, 997
    fishing industry, **4:**498
    GDP in billions of dollars, **8:**1027
    germ warfare, **5:**591
    as gunpowder source, **9:**1190
    human ecology, **5:**649–650, 659
    human rights, **5:**664–665
    hydropower, **5:**682, 685
    irrigation system, **6:**728
    landscape paintings, **4:**463
    major flood, **4:**510
    nuclear capacity, **7:**894
    one-child policy, **4:***485*
    Peking Summer Palace, **10:**1397
    per capital energy consumption, **9:**1179
    rangeland, **8:**1050
    religious nature beliefs, **10:**1383
    Renaissance European trade contacts,
        9:1190
    scarce orchid, **9:***1126*
    silkworm, **1:**37
    temperate river basins, **9:**1199–1201
    Three Gorges Project, **5:**682, 685, **9:**1200,
        1201, 1216–1219
    Tiananmen Square, **5:**664–665
    tree pruning technique, **9:**1254
    tundra, **10:**1278
    urban park, **7:***951*
    wartime dam breaching, **10:**1325
Chincoteague National Wildlife Refuge,
    7:*867*
Chongqing Municipality (China), **9:**1216,
    1219
Chukchi Sea, **5:**594
Cincinnati (Ohio), **3:**296
Cleveland (Ohio), **5:**505, **9:**1242
Climax (Colo.), **8:**1106
Coast Ranges, **9:**1124, *1125*
Colombia
    birthrate decline, **8:**1010
    coffee plantation, **3:***294*
    percent of underweight children, **5:**673
    tropical dry forest, **9:**1256
Colorado
    Boulder, **5:**609, *616*
    cattle shipment centers, **8:**1048
    Climax observing station, **8:**1106
    Denver International Airport, **7:***878*, 879

Denver smog haze, **9:**1142
Dinosaur National Monument, **7:**954
    irrigation system, **6:***732*
    oil shale deposits, **5:**562
    pueblo dwellings, **5:***659*
    rangeland, **8:**1049, *1051*
    toxic metal smelter plant, **9:***1227*
Colorado River, **4:**420, **5:**586, **6:**730,
    9:1200–1201
Columbia (Md.), **5:**609, **9:**1163, **10:**1302
Columbia County (Ore.), **7:***863*
Columbia River Basin, **2:**234–235
Concord (Mass.), **3:**362, **9:**1214, 1215, *1238*,
    1239
Coney Island (N.Y.), **9:***1135*
Congo River basin, **10:**1272, *1273*, 1274
Connecticut
    Lyme disease, **3:**288
    Merritt Parkway, **8:**1103
Coral Sea, **5:**604
Costa Rica
    birthrate decline, **8:**1010
    ecotourism, **1:**113, **9:**1252
    hydropower, **3:***350*
    restoration ecology, **8:**1086–1087, *1089*
    tropical dry forest, **9:**1256, *1258*
Cotton Exchange (Wilmington, N.C.),
    10:1361
County Limerick (Ireland), **8:***1073*
Crawford (Neb.), **5:***598*
Crete, **10:**1330
Cuba, birthrate decline, **8:**1010
Cumberland Road, **8:**1100
Curitiba (Brazil), **10:**1302
Custer County (Neb.), **6:***750*
Cuyahoga River, **5:**606

Danbury (Conn.), **10:***1337*
Danube River, **2:**236–237
Davis Strait (Greenland), **10:**1379
Dead Sea, **6:**731
Delaware, **9:**1146
Delta National Wildlife Refuge, **5:***558*
Denali National Park, **10:***1279*
Denmark, GDP per capita, **8:**1027
Denver (Colo.), hazy smog, **9:**1142
Denver International Airport, **7:***878*, 879
Detroit (Mich.), **3:**296, **7:**950, **9:***1194–1195*
developing countries
    agricultural systems, **9:**1172
    debt-for-nature swap, **2:**246–247
    deforestation, **2:**252, 254–255
    desertification, **3:**271, 273
    drinking water contamination, **5:**631
    electrification benefits, **3:**389
    environmental damage, **6:**705
    environmental risk equity, **4:**460–462
    family planning programs, **4:**482–485,
        8:1010, 1011–1012, 1013–1015,
        1113–1115
    food-borne illness, **3:**281
    housing, **5:**644, 645
    hunger and famine, **5:**672–673
    indigenous peoples, **5:**689–691
    migrations, **6:**825, 826–827
    population growth, **8:**1008–1014, 1041,
        1054, **9:**1178–1179
    rangeland uses, **8:**1050–1051, 1054
    regenerative agriculture, **8:**1109

shanty-town suburbs, **9:**1160
sustainable development, **9:**1178–1179,
    1181
tourism industry, **9:**1248–1249, 1252
UN conferences, **10:**1292, 1293
urban planning, **10:**1309
Devils Tower monument, **7:**953
Dillon Reservoir (Calif.), **9:**1232
Dillon Reservoir, **9:**1232
Dinosaur National Monument, **7:**954
Disney World, **10:**1363
Dodge Fountain Plaza (Detroit), **7:**950
Dominican Republic, government-spon-
    sored housing, **5:***644*
Donora (Pa.), **9:**142
Don River (Toronto), **8:**1004
Duwamish River, **4:**441

Eagle Butte open-pit mine (Wyo.), **7:**840
Eagle Harbor (Wash.), **9:***1168*
Eagle Summit (Alaska), **10:**1278
East China Sea, **9:**1200
Eastern Woodland Complex, **8:**997
East St. Louis (Ill.), **10:**1309
Ecuador
    El Niño, **3:**356, 357
    tropical dry forest, **9:**1256
Egypt
    ancient waste dumps, **10:**1330
    animal domestication, **1:***32*
    Aswan High Dam, **1:**80–81, **5:**639, 682
    demographic transition, **8:**1011
    family planning program, **4:**483
    green revolution, **5:**620
    Hatshepsut Mortuary Temple, **6:***756*
    landscape design, **6:***756*, 757
    pyramids, **10:**1397
    See also Nile basin; Nile River
El Chichon volcano, **2:**185
El Morro monument, **7:**953
El Paso (Tex.), **9:***1230*
El Salvador
    fertility rate, **8:**1010
    geothermal power plant, **8:***1084*
England. *See* Great Britain
Equitable Savings and Loan Building, **1:**68,
    10:*1398*
Erie, Lake, **5:**606
Erie Canal, **1:**74, **2:**145/g
Esfahan (Iran). *See* Isfahan
Ethiopia, **3:**298, **5:**690
    desert locust control, **6:***710*
    El Niño effects, **3:**357
    per capital energy consumption, **9:**1180
    percent of underweight children, **5:**673
    refugees, **6:**825, *827*
Euclid (Ohio), **10:**1399, 1400
Euphrates River and Valley, **2:**144, **5:**600,
    6:751, **9:**1199
Everglades, **2:**145, **4:**468–471, **7:**955, **8:**1020
    migrant workers, **6:***822*
    red tide, **1:***24*, 25

Fairlawn (N.J.), **10:**1308
Fallingwater, **1:***68*, **10:**1390–1391
Farmington (W. Va.), **7:**899
Faroe Islands, **10:***1379*, 1380
Felsenthal National Wildlife Refuge (Ark.),
    6:*696*

Page numbers in **boldface type** indicate full articles on a subject. Page numbers in *italic type* indicate illustrations or other graphics. All numbers preceding colons indicate volume numbers.

Harborplace (Baltimore), **4:***413*, **10:***1361*
Harbur Lothian (Scotland), **9:***1172*
Hawaiian Islands
    biota, **2:**130–131
    carbon cycle, **2:**157
    climate change, **2:**187
    El Niño effects, **3:**357
    Green Party, **5:**619
    oceanography, **7:**907, 910
    pest inspection and quarantine, **7:**964
    whale migration to, **7:**868
    wildflowers, **10:**1385
Hayman Great Barrier Reef Resort,
    9:*1251*–1252
Henan Province (China), **10:**1325
Heracleopolis (Egypt), **10:**1330
Hetch Hetchy Debate, **8:**990–991, 1016,
    1018
Hetch Hetchy Valley, **8:**1018, *1019*
Highway Beautification Act, **4:**465
Hilton Head Island (S.C.), **4:**423
Hokkaido (Japan), **9:***1133*
Homer (Alaska), **4:***434*, **9:***1132*
Homestead (Pa.), **6:***700*
Honduras
    fertility rate, **8:**1010
    hurricane disaster, **7:**857
Hong Kong, demographic transition, **8:**1009
Houston (Tex.), commuter Clean Air re-
    quirements, **9:**1245
Huanghe River (Yellow River), **9:**1199, *1200*
    sediment transport, **9:**1198
    silt in, **9:**1153, 1156–1157, 1218
Huayuankow dike (China), **10:**1325
Hubbard Brook Experimental Forest
    (N.H.), **10:**1370, 1371
Hudson River, **1:**28, **4:**422
Hudson River Valley, **4:**464

Iceland, whaling industry, **10:**1380, 1381
Idaho
National Reactor Testing Station, **4:***423*
Palouse Prairie, **8:**1053
rangeland, **8:**1049, 1053
Rocky Mountains image, **3:***308*
stream viral pollution, **10:***1320*
Ile-de-France (Paris), **5:**656
Illinois
    first U.S. nuclear reactor, **7:**894
    irradiated food marketing, **6:**726
    major flood, **4:**510
    remote sensing information, **8:**1077
    tree-planting buffer zone, **10:**1368
    urban problems, **10:**1309
India
    Ayurveda medicine, **6:**804–805
    Bhopal disaster, **1:**108–109, **2:**205
    biogas digesters, **1:**115
    cyclone, **7:**857
    dengue hemorrhagic fever, **3:**286
    ecolabeling, **3:**314
    family planning program, **2:**223, **4:**483, 484
    fertility rates, **8:**1010
    fuelwood portage, **5:***568*
    GDP in billions of dollars, **8:**1027
    green revolution, **5:**620, 671
    groundwater collection, **5:***623*
    Hinduism, **8:**1069–1070
    Indus Valley, **6:**751, **10:**1330

irrigation systems, **6:**728
Jainism, **8:**1064, *1066*
landscape architecture, **6:**757
per capital energy consumption, **9:**1179
percent of underweight children, **5:**673
Taj Mahal, **6:**757, **9:***1249*, 1253
urban overcrowding, **10:**1303
Indiana, urban problems, **10:**1309
Indian Ocean, **9:**1135
Indonesia
    Bali, **1:***10*
    demographic transition, **8:**1009
    family planning program, **4:***483*
    GDP in billions of dollars, **8:**1027
    Hindu Temple relief sculpture, **6:***806*
    indigenous people, **5:**690
    mining, **9:***1180*
    percent of underweight children, **5:**673
    petroleum, **8:**980
    smog, **9:***1140*
    soil erosion, **9:**1155
Indus Valley, **6:**751, **9:**1199, **10:**1330
Inner Harbor Project (Baltimore). *See*
    Harborplace
Intercoastal Waterway, **4:***425*
Intermountain Bunchgrass, **8:**1053
Iowa, **4:**510, **8:***1113*
Iran
    ancient pottery, **1:***62*
    nomadic Qashqai, **5:***688*
    OPEC membership, **7:**919, 920, 921, 922,
        923
    petroleum, **8:**980
    weaving center, **3:***343*
Iraq, **5:**671
    Bedouin, **2:***259*
    Gulf War, **7:**922, 923
    OPEC membership, **7:**911, 920
    ziggurat, **10:**1396–1397
Ireland
    Benedictine monks, **8:***1073*
    tundra, **10:**1278
Irvine (Calif.), **6:**726, **10:**1302
Isfahan (Esfahan), **4:***343*; **6:**757
Islamabad (Pakistan), **3:**296
Israel
    Arab wars, **7:**920–921
    Ethiopian Jewish refugees, **6:***827*
    kibbutz utopianism, **10:**1317
    soil erosion, **9:**1155
Italy
    dioxin accident, **3:**282–283
    Earth Day activities, **3:**306
    GDP in billions of dollars, **8:**1027
    geothermal plant cooling towers, **3:***390*
    landscape design, **6:***757*, 758
    major floods, **4:**510

Jakarta, **2:***204*
Jamaica, peat fields, **1:***116*
Japan
    air pollution, **1:***18*, *21*
    biotechnology, **2:**135
    Buddhism, **8:**1068
    chrysanthemum reverence, **8:**999
    ecolabeling, **3:**314
    fishing industry, **4:**498
    gardens and landscape design, **4:**463,
        6:*750*, 757, *760*

GDP in billions of dollars, **8:**1027
GDP per capita, **8:**1027
harbors, **9:**1134
Human Genome Project, **5:**653
incineration, **10:**1335
landscape paintings, **4:**463
major flood, **4:**510
Minamata disease, **6:***814*, 815
nuclear capacity, **7:**894
PCB contamination, **7:**958
pesticide spraying, **7:***967*
planned community, **10:**1302
religious nature beliefs, **10:**1383
shoreline, **9:***1133*
shoreline settlement, **9:**1134, 1135
temperate rain forest, **9:**1196
TERRA joint project, **3:**308, **8:**1079
urban projects and new towns, **10:**1303
waste disposal, **10:**1336
water hyacinth specimens, **10:**1362
whaling industry, **10:**1379, 1380, 1381
See also Tokyo
Java (Indonesia)
    family planning, **4:***483*
    soil erosion, **9:**1155
Jericho (Jordan Valley), **8:**996
Jerusalem, **10:**1330
Jiangsu Province (China), **10:**1325
Jordan
    desert bloom, **2:***262*
    soil erosion, **9:**1155
Jordan Valley, **8:**996
Jurong (Singapore), **10:**1302

Kalahari Desert, **6:***824*
Kansas
    cattle shipment centers, **8:**1048
    grasslands, **5:**603
    major flood, **4:**510
Karachi (Pakistan), **3:**296
Katsura Imperial Palace (Kyoto), **6:**757
Kazakhstan, **1:**60
Kentucky
    coal reserves, **2:**197
    groundwater contamination, **5:***622*
    major flood, **4:**510
    TVA mandate in, **9:**1202
Kenya
    carrying capacity limit, **2:**165
    demographic transition, **8:**1009
    Earth Day activities, **3:**307
    ecotourism, **1:**113, **9:**1252
    El Niño effects, **3:**357
    energy-efficient stoves, **9:***1178*
    grasslands, **5:***601*
    green belt, **5:***608*
    green revolution, **5:**620
    irrigation canal, **6:***728*
    IUCN, **6:**733
    UN Environment Programme, **10:**1296
    wildebeests habitat, **7:**866
Khartoum (Sudan) flood, **4:**510
Kiel Canal, **2:**145
Kissimmee River, **2:**145, **8:**1087–1088
Kista (Sweden), **10:**1302
Kodiak Island, **10:**1379
Korea
    demilitarized zone wildlife sanctuary,
        10:*1324*

Page numbers in **boldface type** indicate full articles on a subject. Page numbers in *italic type* indicate illustrations or other graphics. All numbers preceding colons indicate volume numbers.

Page numbers in **boldface type** indicate full articles on a subject. Page numbers in *italic type* indicate illustrations or other graphics. All numbers preceding colons indicate volume numbers.

Index of Persons

Page numbers in **boldface type** indicate full articles on a subject. Page numbers in *italic type* indicate illustrations or other graphics. All numbers preceding colons indicate volume numbers.

Bradshaw, Thornton, **8:**1065
Brady, Lynn R., *as contributor,* **6:**798–803
Breuer, Marcel, **1:**68
Brezonik, Patrick L., *as contributor,* **9:**1128–1131
Briggs, Shirley A., *as contributor,* **2:**166–167
Brooks, Cynthia, *as contributor,* **2:**138–139
Browder, Joe, *as contributor,* **4:**468–471
Brower, David, **2:**215, **4:**443, 447
Brown, Capability, **6:**758
Brown, Lester R., **6:**780
Brown, Peter G., *as contributor,* **4:**460–462
Browner, Carol, **4:**451
Brueghel, Pieter the Elder, **10:***1396*
Brundtland, Gro Harlem, **9:**1176
Bryant, William Cullen, **2:**214
Buckland, William, **6:**784
Buffler, Patricia A., *as contributor,* **4:**455–459
Burmeister, Leon F., *as contributor,* **5:**636–637
Burroughs, John, **4:**422, **6:**775–776, **8:**1110
Burton, Philip J., *as contributor,* **4:**520–525
Bush, George, **4:***431,* 451, 496
Byrd, Richard E., **1:**45
Byrd, Warren T., Jr., *as contributor,* **6:**756–761

Cairns, John, Jr., *as contributor,* **8:**1086–1089
Caldwell, Lynton Keith, *as contributor,* **5:**664–667
Callenbach, Ernest, **10:**1317
Calvin, Melvin, **8:**987
Cannon, Walter B., **5:**575, **6:**806
Carey, John, *as contributor,* **5:**646–657
Carlyle, Thomas, **3:**362
Carr, Archie, **2:**166
Carson, Rachel, **2:**136, **166–167,** 240, **4:**432, **5:**647, **6:**709, 737, 738, 778, *779*
    environmental movement and, **4:**422, 442
    excerpt from *Silent Spring,* **2:**241
    Fish and Wildlife Service post, **4:**494
    as pesticide control impetus, **1:**12
Carter, Jimmy, **4:**449, 496, **6:**783
Castells, Manuel, **6:**811
Catlin, George, **2:**214
Caudill, Harry, **1:**204, **2:**204
Cavanagh, Ralph, *as contributor,* **8:**1034–1037
Cavendish, Henry, **5:**677
Chakrabarty, Anada, **7:***915*
Charlemagne, **2:**212
Christy, Henry, **10:**1286
Church, Frederick Edwin, **4:**464
Church, Thomas, **6:**759
Clark, Kenneth, **4:**463
Clark, Merriweather, **2:**140
Clausen, Don, **4:***438*
Clay, Jason W., *as contributor,* **5:**688–691
Clements, Frederick W., **2:**215, **8:**1054
Cleveland, Horace W. S., **6:**756, 758, **7:**949
Cleveland, L. R., **9:**1185
Clinton, Bill, **2:**141, **4:**451, **5:***667,* **7:**954, **8:***1032,* **9:***1233*
Cohen, Stanley, **5:**653
Cohn, Susan, *as contributor,* **3:**364–369
Cole, David E., *as contributor,* **1:**90–93, **10:**1392–1395
Cole, Thomas, **2:**214, **4:**464
Coleridge, Samuel Taylor, **3:**362, **9:**1239

Collor de Mello, Fernando, **4:***429*
Commoner, Barry, **3:**306, **5:**619, **6:**778
Cook, James, **5:**604–605
Cooper, James Fenimore, **2:**214, **4:**464, **6:***774,* 775
Cooper-Janis, Maria, **8:**1066
Coppock, Robert, *as contributor,* **2:**184–189, **5:**612–617
Corbusier, Le, **1:**67
Cormack, Malcolm, *as contributor,* **10:**1284—1285
Costle, Douglas, **4:**449
Cottrill, Ken, *as contributor,* **7:**838–841
Cousins, Norman, **6:**806
Cousteau, Jacques, **2:230–231**
Covello, Vincent T., *as contributor,* 1090–1095
Cowles, Henry, **2:**215
Crabtree, Pam J., *as contributor,* **1:**62–64
Cramer, Craig, *as contributor,* **8:**1108–1109
Crichton, Michael, **3:**293
Crick, Francis, **3:**292, **5:**577, 653
Crowe, Sylvia, **6:**759
Cummings, Ronald G., *as contributor,* **2:**220–221
Cummins, Kenneth W., *as contributor,* **8:**1098–1099
Curie, Marie and Pierre, **8:**1046
Curtis, Dale, *as contributor,* **10:**1310–1315
Cuvier, Georges, **5:**582, 583, **7:**942

Dai Qing, **9:**1219
Dalai Lama, **8:**1065
Darby, Abraham, **9:**1191
Darling, Frank Fraser, **3:**361
Darling, Jay ("Ding"), **4:**494
Darmstadter, Joel, *as contributor,* **3:**402–403
Darwin, Charles, **1:**48, **2:**227, **238–239,** **5:**577–578, 583, **6:**740, 741, 812, **9:**1164
    evolution theory, **4:**472, 473, 476–477
    on Galápagos finches, **7:**875, 877
    Lyell's views and, **6:**784, 785
    Wallace's evolution theory and, **10:**1322, 1323
Darwin, Robert, **2:**238
Davies, Terry, *as contributor,* **4:**448–452
Day, Gordon M., **4:**493
Deisenhofer, Johann, **8:**987
Deisler, Paul F., Jr., *as contributor,* **7:**971–975
Delwiche, C. C., *as contributor,* **7:**883–889
Democritus, **3:**354
Denison, Edward F., *as contributor,* **8:**1026–1027
Devoto, Bernard, **6:**777
De Vries, Hugo, **4:**473
Dewey, John, **5:**649, 650
Dickinson, Emily, **9:**1236
Dickson, L., *as contributor,* **10:**1350–1354
Dillard, Annie, **6:**781
Dingell, John D., **4:**451
Dorfman, Robert, *as contributor,* **1:**104–105
Douglas, Marjory Stoneman, **8:**1020
Douglas, William O., **4:**465
Downing, Andrew Jackson, **6:**758, **7:**948–949
Doxiades, Constantinos, **3:296–297**
Doyle, Robert D., *as contributor,* **10:**1350–1354

Drake, Colonel, **8:**980
Drake, Elisabeth, *as contributor,* **3:**376–381
Drake, James A., *as contributor,* **4:**478–481
Droll, Richard, **3:**277
Dubos, René, **2:**131, **3:302–305,** **4:**468, 493, **5:**575, **6:**778, 797, **10:**1326, 1397
    antibiotics and, **1:**54–55
    as contributor, **1:**6–7, **3:**404–407, **4:**502–507, **5:**572–575, 656–661, **7:**870–873, 956–957, **10:**1382–1383
    Earth Day and, **3:**306
    ecosystem case histories by, **3:**347
    on hedgerow biotic community, **2:**131
    humanized landscape and, **4:***506*
    "Think Globally, Act Locally" saying, **9:**1212
    UN Conference on the Human Environment and, **10:**1290, 1328
Duncan, Norman E., *as contributor,* **7:**858–863, 918–923
Dunlap, Riley E., *as contributor,* **4:**436–441, 442–447
Duxbury, Dana, *as contributor,* **5:**624–625

Eaubonne, Francoise d', **3:**312
Eblen, Ruth A., *as contributor,* **3:**302–305, 354–355, **4:**434–435, 502–507, **5:**572–575, **6:**786–787, **7:**914–915, 956–957, **9:**1212–1213, 1232–1233, 1234–1235, **10:**1290–1293, 1295–1297, 1298–1301, 1316–1317, 1388–1389
Eblen, William R., *as contributor,* **2:**136–137, **3:**342–347, 404–407, **4:**416–419, **5:**656–661, **6:**820, **7:**870–873, 875–877, **8:**1046–1047, **9:**1158–1159, 1206–1207, **10:**1363, 1382–1383
Eckbo, Garrett, **6:**759
Eckholm, Erik, **10:**1329
Edison, Thomas, **3:**348, 376, 385, 387
Edmonson, W. T., *as contributor,* **8:**992–995
Ehlers, Jake, *as contributor,* **7:**844–845, **8:**1110–1111
Ehrlich, Gretel, 781
Ehrlich, Paul, **2:**216, **3:**306, **6:**778
Einstein, Albert, **7:**892
Eiseley, Loren, **6:**779
Eisenhower, Dwight, **4:**496
Eliot, Charles, **7:**949
Ellis, Gary B., *as contributor,* **2:**132—133
Ellsaesser, Hugh W., *as contributor,* **1:**82–87
Elton, Charles S., **3:**360–361, **4:**481
Emerson, Ralph Waldo, **2:**214, **3:362–363,** **6:**709, **9:***1237*
    as influence on Thoreau, **3:**362, **6:**775, **9:**1214
    literature by, **6:**775
    Muir compared with, **7:**843
    on nature as key to self-discovery, **7:**952
    oration on *The Method of Nature,* **3:**363
    text from *Nature,* **2:**219
    as transcendentalist, **9:**1236, 1237, 1239
Empedocles, **3:**354
Enslow, Lyn, **10:**1386
Erikson, Kai, **2:**204
Esterman, Pamela R., *as contributor,* **1:**26–31
Evans, Brock, **6:**745
Evensky, Jerry, *as contributor,* **9:**1138–1139

Page numbers in **boldface type** indicate full articles on a subject. Page numbers in *italic type* indicate illustrations or other graphics. All numbers preceding colons indicate volume numbers.

Page numbers in **boldface type** indicate full articles on a subject. Page numbers in *italic type* indicate illustrations or other graphics. All numbers preceding colons indicate volume numbers.

Comprehensive Index

Page numbers in **boldface type** indicate full articles on a subject. Page numbers in *italic type* indicate illustrations or other graphics. All numbers preceding colons indicate volume numbers.

Page numbers in **boldface type** indicate full articles on a subject. Page numbers in *italic type* indicate
illustrations or other graphics. All numbers preceding colons indicate volume numbers.

coal, **2:**199
   combustion residues, **10:**1347, 1348
   incineration, **5:**687
Asheville (N.C.), **7:**917
asphalt, **8:**1101
aspirin, **8:**998
Associated General Contractors, **7:**903
Association of Schools of Public Health, **8:**1030
Association of Southeast Asian Nations, **4:**427
asthma, **5:**631, 633
astrogeology, **5:**584
astrophysics, **8:**1106–1107
Aswan High Dam, **1:**80–81, **5:**639, 682
AT&T Matrix System, **6:**698, *699*
Athens (Greece)
   ancient waste disposal, **10:**1330
   modern, **3:***297*
   Temple of Athena Nike, **1:**65–66, *65*
Atlanta (Ga.), **9:**1242, 1246
Atlantic Ocean, **10:**1358
   tidal power, **3:**395
   U.S. fishing industry, **4:**500, 501
Atlas, Ronald M., *as contributor,* **1:**94–99
atmosphere, **1:**82–87
   acid rain, **1:**2, 87
   air pollution, **1:**18–23
   Arctic, **1:***72,* 73
   biogeochemical cycling, **1:**96–97
   carbon cycle, **1:**82, 84–85, **2:**154–155, 156–159, **5:**616
   carbon dioxide level, **1:**86–87, **2:**152, 157, 158–159, 169, 185, 187
   carbon monoxide concentrations, **1:**260
   CFCs in, **2:**168–169
   climate system, **2:**184–189
   composition, **1:**82–84, 87, **5:**573
   cosmic radiation, **8:**1043–1044
   dissolved oxygen from, **3:**290
   Earth Observing System, **3:**309–310
   elements, **3:**354–355
   El Niño, **7:**912
   environmental law, **4:**427–428
   Gaia hypothesis, **5:**573–574
   greenhouse effect, **5:**612–617, **6:**705
   heat from, **5:**555
   hydrogen, **5:**677–678
   layers, **1:**82, *83*
   meteorology, **6:**816–819
   methane, **7:**861
   mineral cycle, **6:**828, 829, 830, 833
   nitrogen, **7:**880–882, 883
   nitrogen cycle, **7:**880–881, 883–889
   origin of life, **4:**475
   outer space, **7:**926
   oxygen, **7:**928–931, 932
   oxygen cycle, **7:**932–937
   ozone layer, **7:**933, 938–943, 975
   ozone layer depletion, **4:**427–428, 438, **6:**704, 705, **8:**1040
   photosynthesis, **8:**986–989
   smog, **9:**1140–1143
   trace metals emissions, **9:***1227*
   water cycle, **3:**301, **10:**1353, 1355, *1357, 1358*
   See also air pollution; carbon cycle; greenhouse effect

atom
   element composition, **3:**354, 355
   energy, **3:**377
   fission, **7:**892, 893–893
   hydrocarbon, **7:**972
   hydrogen, **5:**677, 678
   nitrogen, **7:***881*
   oxygen, **7:***928,* 929, 932, 934–935, 938
atomic bomb. *See* nuclear weapons
Atomic Energy Act, **7:**893, 899
Atomic Energy Commission, **4:***423,* **7:**893, 894, 897, **10:**1387
ATP, **1:**94
Auchter, Thorne, **7:**902
Audubon, John James, **1:**88–89, **2:**214, *215,* **6:**709, 775
Audubon Society. *See* National Audubon Society
Augustine, Saint, **8:**1074, **10:**1317
Austin (Tex.), 1385
Austin, Mary, **6:**776–777
Australia
   Aborigines, **8:**1062, *1063*
   agricultural technology, **9:**1192
   chaparral, **2:**171
   El Niño effects, **3:**357
   GDP per capita, **8:**1027
   grasslands, **1:**123, **5:**598
   Great Barrier Reef, **5:**604–605, **9:**1251–1252
   Hayman Island Resort, **9:**1251–1252
   ranches, **8:**1048
   rangeland, **8:**1050, 1053, 1054
   temperate rain forest, **9:**1196
   tropical dry forest, **9:**1256
   unique biota, **2:**130, *131*
Australian Aborigines, **8:**1062, *1063*
Australian Rangeland Society, **8:**1050
Australopithecus, **1:**50
Austria, **5:**619, **9:**1246
Autobahn, **8:**1103
automobile, **1:**90–93
   alternative fuels, **1:**91–92, 117, **5:**555, 557, 561, 679, **7:**863
   batteries, **3:**383, 384
   Btu tax, **3:**403
   commuter use, **9:**1244, 1245
   driving risk, **8:***1092,* 1093
   electric cars, **3:***384,* 388, **5:***557, 564, 676*
   emissions, **2:**160, *169,* 178, 179, **8:**982, 1104–1105, **9:**1140, 1141–1142
   energy use, **3:**377, 380
   fuel economy standards, **5:**566–567
   fuel energy efficiency, **8:**981, **9:**1180
   gasoline, **7:**975
   industrialization and, **6:**702–703
   motor oil disposal, **5:**624–625
   recreational use, **8:**1057
   recycling and, **3:**347, **6:**697
   roads and highways, **8:**1101–1105, **9:**1240–1241, *1243*
   rubber use, **9:**1193
   smog from, **9:**1140, 1141–1142
   suburbs and, **9:**1161–1162, 1242
   urban traffic, **10:**1302–1304, 1308, 1309
   zero-emission, **10:**1392–1395
avalanche, **5:**595–596
Avery, Oswald T., **3:**303–304, **5:**577

Ayres, R. U., *as contributor,* **6:**706–707
Ayurveda medicine, **6:**804–805

Babbitt, Bruce, **7:**954
baboon, **1:**49, 51
Babylon, **6:**757, **8:**1100, **10:**1397
Bachman, John, **1:**89
Bacon, Francis, **6:**708
bacteria, **1:**94–99, **7:***934*
   antibiotic, **1:**54–55, 114, **3:**302, 303–304
   biodegradation by, **1:**98–99, 115, **2:**210, 7:959
   biotechnological uses, **1:**98, 113
   cloning, **2:**190
   compost stabilization, **10:**1339
   diseases, **3:**284, *286,* 288–289
   dissolved oxygen and, **3:**291
   in ecosystem, **3:**343
   epidemiology, **4:**455, *456*
   eutrophication, **8:***985,* 995
   evolution, **4:**475, 477
   food chain, **9:**1129, 1130
   food contamination, **3:**276, 279–280, **9:**1122–1123
   food irradiation against, **6:**722–727
   gene splicing, **5:**579
   genetically engineered, **2:**134
   methane production, **7:**861
   military use, **5:**590–591, **10:**1325
   nitrate-to-nitrite conversion, **7:**882
   nitrogen cycle, **7:**881, *883,* 884–885, 887, **10:**1374
   oil-spill eating, **7:**915
   oxygen cycle, **7:**936, 937
   as parasites, **7:**944
   PCB biodegradation by, **7:**959
   as pest control, **7:**963, 964–965
   phosphorus cycle, **8:**984–985
   photosynthesis, **8:**986–989
   plankton, **8:**993–994, 995
   salmonella, **3:**279, **6:**722, *725,* 726, **9:**1122–1123
   symbiosis, **9:**1182–1185, 1204–1205
   tuberculosis, **10:**1276
   water-borne pollutants, **10:**1368, 1386
bacteriological warfare, **5:**590–591, **10:**1325
bacterioplankton, **8:**993
bacteroids, **9:**1182–1183
Bahamas, **9:***1130*
Baikal, Lake, **1:**100–101, **9:**1129
Baker, Victor R., *as contributor,* **9:**1198–1201
bald eagle, **3:***372, 375*
Bali rice paddies, **1:***10*
Ballinger, Richard, **8:**990
Ballinger-Pinchot Controversy (1910), 8.990. *See also* preservation vs. conservation
Baltic Sea, **2:**243
Baltimore (Md.)
   commuter Clean Air requirements, **9:**1245
   rapid transit system, **9:**1242
   waterfront development, **4:***414,* **10:**1303, *1361*
bamboo, **1:**102–103
Bancroft, John M., *as contributor,* **2:**260–265
Bangladesh
   bamboo, **1:***103*
   family planning program, **8:**1010

Page numbers in **boldface type** indicate full articles on a subject. Page numbers in *italic type* indicate illustrations or other graphics. All numbers preceding colons indicate volume numbers.

as greenhouse gas, **2:**185, 188, **5:**612, 615, 616

as natural gas product, **7:**861, 862

oxygen cycle and, **7:**929, 930, 932, *933*, 937

percent in earth's atmosphere, **5:**573

petroleum emissions, **8:**982, 983

sources, storage, and level, **1:**84–85

See also greenhouse effect

carbon monoxide, **2:160–161**

air pollution, **1:**18, 19, *20*, 21, 22, 23, **2:**160

Clean Air Act, **2:**178

emissions, **2:**160, 178, **8:**982

emissions control, **1:**92

poisoning symptoms, **2:**161

carbonyl sulfide, **1:**86

carcinogen, **2:**146, **162–163**, **4:**548, **5:**630, 631, 632, 634

benzene, **7:**973

bioassay, **1:**110–111

chemical, **2:**162, 173, 174, **3:**279

chemical mutation, **9:**1230

nitrate/nitrite, **7:**882

occupational, **7:**900

radon as, **8:**1046

ultraviolet radiation, **7:**938

careers, environment-related. *See* employment and careers

Carey, John, *as contributor*, **5:**646–657

Caribbean region, **9:***1130*

dengue hemorrhagic fever, **3:**287

desalination plants, **2:**258

hurricanes, **7:**913

Carlyle, Thomas, **3:**362

carnivores, **3:**316, 361, **4:**516, 522

carotenoids, **3:**278

carpal tunnel syndrome, **7:**900

Carr, Archie, **2:**166

carrying capacity, **2:164–165**, 217, **3:**331, **8:**1008–1015

Carson, Rachel, **2:**136, **166–167**, 240, **4:**432, **5:**647, **6:**709, 737, 738, 778, *779*

environmental movement and, **4:**422, 442

excerpt from *Silent Spring*, **2:**241

Fish and Wildlife Service post, **4:**494

as pesticide control impetus, **1:**12

Cartago (Costa Rica), **3:***350*

Carter, Jimmy, **4:**449, 496, **6:**783

Casa Grande (Arizona), **1:***42*

Cascade Mountains, **7:**950, **8:**1053, **9:***1170*

Cascades National Park, **7:***952*

Cascadian Farm (Wash.), **9:***1170*

Caspian Sea, **2:**145

Castells, Manuel, **6:**811

cat, **1:**35

Çatal Hüyük (Turkey), **1:**34

catalytic converter, **1:**92–93, **4:***462*

catalytic cracking process, **1:**173

cataracts, **2:**168, **10:**1288

caterpillar, **6:***717*

Catholicism, **10:**1329

Catlin, George, **2:**214

cattle. *See* animal, domesticated; ranching

Caudill, Harry, **1:**204, **2:**204

Cavanagh, Ralph, *as contributor*, **8:**1034–1037

Cavendish, Henry, **5:**677

cave painting, **6:**708, *709*

Celebrations of Life (Dubos), **9:**1212

cellulose, **9:**1185

Celtic Christianity, **8:**1074, 1075

Centers for Disease Control, U.S., **7:**958, **8:**1030

Central Arabian Arch aquifers, **6:***730*

Central Park (N.Y.C.), **4:**464, **6:**758, 761, **7:**916, *917*, 949, 953

CFC, **2:168–169**

carbon cycle, **2:**159

global warming, **2:**185, **9:**1250

greenhouse effect, **5:**612, 617

industrialization and, **6:**703–704, 705

ozone layer thinning, **7:**939–941, **8:**1041, **10:**1288

phaseout timetable, **3:**387

pollution trading program, **8:**1007

taxation of, **3:**338

chaco, **1:**119

Chad, **8:**1011

Chagas's disease, **7:**947

chain reaction, nuclear, **7:**893–894

Chakrabarty, Anada, **7:***915*

Challenger, HMS, **7:**905, 910

Ch'ang (Yangtze) Delta, **8:**996, 997

Ch'ang River (Yangtze River), **9:**1199–1200, 1201, 1216, *1217*, 1218, 1219, **10:**1324

Chantilly (France), **6:**757

chaparral, **1:**119, *123*, **2:170–171**, **8:**1053

charcoal, **2:**153, 173

Charlemagne, **2:**212

Charles River Natural Valley Storage Project, **10:**1372–1373

Charleston (S.C.), **4:**464

Charleston Museum (S.C.), **7:**846

Chattanooga (Tenn.), **9:**1246

chemical energy. *See* energy, chemical

chemical industry, **2:172–175**

agricultural hazards, **5:**633–634, 636–637

agricultural products, **1:**12–13, **6:**704, 710–711

air polluting off-gassing, **1:**18

Bhopal disaster, **1:**108–109, **2:**205

carcinogens, **2:**162, 173, 174, **3:**279, **9:**1230

conifer forest sources for, **4:**525

drinking water purification, **3:**298

emissions controls, **2:**179

environment careers in, **3:**365–366

fertilizer, **4:**487

food contamination, **3:**276, 279, 280–281

Great Lakes discharges, **5:**606

hazardous wastes dumping, **4:**439

health hazards, **5:**631, 632, 633

industrialization and, **6:**701–705

insecticides, **6:**710–711, 717

Love Canal toxic waste, **6:**782–783

plant explosion, **2:***173*

pollution by, **6:**705

Superfund and, **9:**1169

synergism, **9:**1187

toxicology, **9:**1228–1231

toxic wastes, **4:**439, **5:**627, *629*, 631

water contamination, **3:**277

See also herbicide; insecticide; pesticide

Chemical Manufacturing Association, **1:**109

chemical notation, **7:**935

chemical oceanography, **7:**910–911

chemolithotrophic bacteria, **1:**94, 96, 97

chemotherapy, **2:**148

Chengdu (China), **9:**1200

Chequamegon National Forest, **8:***1058*

Chernobyl, **1:**73, **2:176–177**, *176–77*, 205, **6:**705, **7:**896, **8:**1046, **10:**1326

Chesapeake and Ohio Canal, **1:**74

Chesapeake Bay, **2:**242, **3:**275, **6:**830, **10:**1368

chewing tobacco, **9:**1222

Chiapas (Mexico), **5:***666*

Chicago (Ill.)

commuter Clean Air requirements, **9:**1245

garden suburb, **7:**917, **10:**1307

park system, **7:**949

rapid transit system, **9:**1242

slaughterhouses, **8:**1048

Chicago World's Fair, **7:**917

chicken, **1:**35

Chihuahuan Desert, **8:**1053

child nutrition, **5:**671–672, *673*

Children's Fund, UN, **10:***1296*, 1301

Chile

birthrate decline, **8:**1010

fishing industry, **4:**498

IUCN, **6:**733

temperate rain forest, **9:**1196

chimpanzee, **1:***48*, 49, 50, 51

China

alternative medicine, **6:***802*, 804, 805

ancient waste disposal, **10:**1331

aquaculture, **1:**56

bamboo, **1:**103

biogas digesters, **1:**115

Bronze Age weapons, **9:***1188*

Buddhism, **8:**1068

chrysanthemum reverence, **8:**999

Communist planned community, **10:**1317

demographic transition, **8:**1009

dengue hemorrhagic fever, **3:**286

domesticated animals, **1:**35–36

double cropping, **9:**1171

Earth Day activities, **3:**307

earthquake, **7:**856

erosion rates, **9:**1153, 1156–1157

family planning program, **4:**483, 484, *485*, **8:***1010*

feng shui (geomancy), **5:**588–589, **6:***749*

first plant domestication, **8:**996, 997

fishing industry, **4:**498

GDP in billions of dollars, **8:**1027

germ warfare, **5:**591

as gunpowder source, **9:**1190

human ecology, **5:**649–650, 659

human rights, **5:**664–665

hydropower, **5:**682, 685

irrigation system, **6:**728

landscape paintings, **4:**463

major flood, **4:**510

nuclear capacity, **7:**894

one-child policy, **4:***485*

Peking Summer Palace, **10:**1397

per capital energy consumption, **9:**1179

rangeland, **8:**1050

religious nature beliefs, **10:**1383

Renaissance European trade contacts, **9:**1190

scarce orchid, **9:***1126*

silkworm, **1:**37

Page numbers in **boldface type** indicate full articles on a subject. Page numbers in *italic type* indicate illustrations or other graphics. All numbers preceding colons indicate volume numbers.

Page numbers in **boldface type** indicate full articles on a subject. Page numbers in *italic type* indicate illustrations or other graphics. All numbers preceding colons indicate volume numbers.

detergent, **3:274–275**
  ecolabeling, 313
  phosphorus in, **8:**984, 985
  plastic packaging of, **8:**1003
Detroit (Mich.), **3:**296, **7:**950, **9:***1194–1195*
deuterium, **7:**894
developing countries
  agricultural systems, **9:**1172
  air pollution, **9:***1140*
  debt-for-nature swap, **2:**246–247
  deforestation, **2:**252, 254–255
  desertification, **3:**271, 273
  drinking water contamination, **5:**631
  electrification benefits, **3:**389
  environmental damage, **6:**705
  environmental risk equity, **4:**460–462
  family planning programs, **4:**482–485,
    8:1010, 1011–1012, 1013–1015,
    1113–1115
  food-borne illness, **3:**281
  housing, **5:**644, 645
  hunger and famine, **5:**672–673
  indigenous peoples, **5:**689–691
  migrations, **6:**825, 826–827
  population growth, **8:**1008–1014, 1041,
    1054, **9:**1178–1179
  rangeland uses, **8:**1050–1051, 1054
  regenerative agriculture, **8:**1109
  shanty-town suburbs, **9:**1160
  sustainable development, **9:**1178–1179,
    1181
  tourism industry, **9:**1248–1249, 1252
  UN conferences, **10:**1292, 1293
  urban planning, **10:**1309
Development Programme, UN,
    10:1294–1295, 1301, 1388–1389
Devils Tower monument, **7:**953
Devoto, Bernard, **6:**777
De Vries, Hugo, **4:**473
dewdrops, **7:**867
Dewey, John, **5:**649, 650
diamonds, **2:**152, 156
Dickinson, Emily, **9:**1236
Dickson, L., *as contributor,* **10:**1350–1354
dieldrin, **5:**606
diesel engines, **2:**201
diet and health, **3:276–281**
  agriculture and, **1:**8
  arthropod delicacies, **6:**715
  cancer and, **2:**147, *148,* 162–163, **3:**277,
    278, 279, 280
  Eskimo, **10:**1283
  food crops, **8:**997
  food irradiation, **6:**724–727
  genetically engineered produce, **2:***135*
  high-fiber benefits, **3:**277
  human evolution and, **1:**50
  human social development and, **1:**52–53
  lead contamination, **6:**766
  mercury risk, **6:**814–815
  socioeconomic status and, **5:**632
  undernutrition programs, **5:**671–672
  UN program, **10:**1295–1296
  vegetarianism, **1:***40*
Dillard, Annie, **6:**781
Dillon Reservoir (Calif.), **9:**1232
dimethyl sulfide (DMS), **1:**25, **6:**833
Dingell, John D., **4:**451
Dinosaur National Monument, **7:**954

dinosaurs, **7:***942*
dioxin, **3:**280, **282–283**, **4:**548, **5:**687
Directory of Environmental Activities and
    Resources in the North American
    Religious Community, **8:**1066–1067
disaster. *See* natural disaster
discrimination, housing, **5:**645
disease. *See* disease, vector-borne; health and
    disease; public health; *specific diseases*
disease, vector-borne, **3:284–289**, **5:**630
  food irradiation against, **6:**727
  germ warfare, **5:**590–591
  as hydropower effect, **5:**684–685
  insecticide use, **6:**710
  insects and, **6:**717
  parasitic, **7:**945, 946–947, 960–961
  virus, **10:**1321
  water-borne bacteria, **10:**1368, 1386
  See also malaria
Disney World, **10:**1363
dispute resolution. *See* alternative dispute
    resolution
dissolved oxygen, **3:290–291**
  lake, **6:**746–747
division of labor, **1:**51
DNA, **3:292–293**, **5:**578–579, **7:**881
  carcinogenic effects on, **2:**146
  chemical mutation, **9:**1230–1231
  cloning, **2:**190, 191, 192, 193
  code, **5:**653
  evolution and, **4:**475
  food irradiation and, **6:**723, 724
  genetic code, **2:**133–134
  Human Genome Project, **5:**652–655
  phosphorus and, **8:**984
  plant domestication analysis, **8:**997
  recombinant technology, **2:**132–133
  structure, **5:**577, 653
  viral, **10:**1318
Doctrine of the Two Contraries, The
    (Empedocles), **3:**354
Dodge Fountain Plaza (Detroit), **7:**950
dodo (bird), **1:**114
dog, **1:**32–33
Dolly (cloned sheep), **2:***190*, 191, 192, 193
domestication, **3:**270, **294–295**
  animal rights, **1:**38–41
  See also animal, domesticated; plant,
    domesticated
Dominican Republic, government-sponsored
    housing, **5:***644*
donkey, **1:**34
Donora (Pa.), **9:**142
Don River (Toronto), **8:**1004
Dorfman, Robert, *as contributor,* **1:**104–105
Douglas, Marjory Stoneman, **8:**1020
Douglas, William O., **4:**465
Downing, Andrew Jackson, **6:**758,
    7:948–949
Doxiades, Constantinos, **3:296–297**
Doyle, Robert D., *as contributor,* **10:**1350–1354
drainage area. *See* watershed
Drake, Colonel, **8:**980
Drake, Elisabeth, *as contributor,* **3:**376–381
Drake, James A., *as contributor,* **4:**478–481
drinking water, **3:298–299**, **9:**1164, **10:**1354
  acid rain and, **1:**3
  contaminants, **1:**99, **3:**277, 298, 299,
    5:631–632, 634, **10:**1364, 1367

corrosion of, **1:**3
Danube River basin, **2:**236
deep well injection, **2:**251
desalination, **2:**258, 259
groundwater as source, **5:**622
nitrate content, **7:**882, 889
quality controls, **2:**180, 181
radon contamination, **8:**1047
safety engineering, **10:**1386
scarcity, **9:**1127
Droll, Richard, **3:**277
dromedary, **1:**34
drought, **1:**106, **2:**158, **3:300–301**, **7:***855*
  agricultural, **1:**11
  categories, **3:**301
  climate change and, **2:**188
  conservation movement and, **2:**217
  desert, **2:**262–263
  desertification, **3:**270, 273
  as El Niño effect, **3:**356–357
  See also Dust Bowl (1930s)
drugs. *See* antibiotic; medicinal drug
Druidism, **8:**1074
dryland. *See* desertification
Dubos, René, **3:302–305**, **4:**468, 493, **5:**575,
    6:778, 797, **10:**1326, 1397
  antibiotics and, **1:**54–55
  as contributor, **1:**6–7, **3:**404–407,
    4:502–507, **5:**572–575, 656–661,
    7:870–873, 956–957, **10:**1382–1383
  Earth Day and, **3:**306
  ecosystem case histories by, **3:**347
  on hedgerow biotic community, **2:**131
  humanized landscape and, **4:***506*
  "Think Globally, Act Locally" saying,
    9:1212
  UN Conference on the Human
    Environment and, **10:**1290, 1328
duck, **1:**35–36
Duck Stamp Act, **7:**852
Duncan, Norman E., *as contributor,*
    7:858–863, 918–923
dunes, **2:***260–261*, 262, 263
  desertification, **3:**272
Dunlap, Riley E., *as contributor,* **4:**436–441,
    442–447
Dust Bowl (1930s), **1:**11, 106, **3:**273, 301,
    5:599, 603, **6:**769, **9:**1149, 1157
Duwamish River, **4:**441
Duxbury, Dana, *as contributor,* **5:**624–625
dyes, **2:**173, 174
dysentery, **10:**1368
dystopianism, **10:**1317
dystrophic biome, **1:**121

Eagle Butte open-pit mine (Wyo.), **7:**840
Eagle Harbor (Wash.), **9:***1168*
Eagle Protection Act, **8:**1021
Eagle Summit (Alaska), **10:**1278
earth
  hydrosphere, **7:**932, 934–935
  land and sea topography, **7:***908–909*, 911
  outer space, **7:**926–927
  See also atmosphere; biosphere; ecosystem
Earth Charter, **8:**1066
Earth Day, **2:**216, **3:306–307**, 313, 314,
    4:436, 437, 443, 444, 447, **6:**781,
    8:1076
  international, **6:**797

Page numbers in **boldface type** indicate full articles on a subject. Page numbers in *italic type* indicate illustrations or other graphics. All numbers preceding colons indicate volume numbers.

Page numbers in **boldface type** indicate full articles on a subject. Page numbers in *italic type* indicate
illustrations or other graphics. All numbers preceding colons indicate volume numbers.

Page numbers in **boldface type** indicate full articles on a subject. Page numbers in *italic type* indicate
illustrations or other graphics. All numbers preceding colons indicate volume numbers.

Page numbers in **boldface type** indicate full articles on a subject. Page numbers in *italic type* indicate illustrations or other graphics. All numbers preceding colons indicate volume numbers.

Page numbers in **boldface type** indicate full articles on a subject. Page numbers in *italic type* indicate illustrations or other graphics. All numbers preceding colons indicate volume numbers.

India (cont'd)
  dengue hemorrhagic fever, **3:**286
  ecolabeling, **3:**314
  family planning program, **2:***223,* **4:***483,* 484
  fertility rates, **8:**1010
  fuelwood portage, **5:***568*
  GDP in billions of dollars, **8:**1027
  Green Revolution, **5:**620, 671
  groundwater collection, **5:***623*
  Hinduism, **8:**1069–1070
  Indus Valley, **6:**751, **10:**1330
  irrigation systems, **6:**728
  Jainism, **8:**1064, *1066*
  landscape architecture, **6:**757
  Orissa, **5:***568,* **7:**857
  per capital energy consumption, **9:**1179
  percent of underweight children, **5:**673
  Taj Mahal, **6:**757, **9:***1249,* 1253
  urban overcrowding, **10:**1303
Indiana, urban problems, **10:**1309
Indian Ocean, **9:**1135
indigenous peoples, **5:688–691**
  coniferous forests uses, **4:**524
  fire uses, **4:**492–493, 528, 545
  fuelwood, **5:***568,* 569
  Great Barrier Reef, **5:**504
  land ownership views, **5:**689, **6:**750–751
  medicine, **6:**800–801, 805
  rangeland use, **8:**1050, *1052,* 1053
  religion, **8:**1062–1063, 1065, 1074–1075
  tundra, **10:**1282–1283
Indonesia
  Bali, **1:***10*
  demographic transition, **8:**1009
  family planning program, **4:***483*
  GDP in billions of dollars, **8:**1027
  Hindu Temple relief sculpture, **6:***806*
  indigenous people, **5:**690
  mining, **9:***1180*
  percent of underweight children, **5:**673
  petroleum, **8:**980
  smog, **9:***1140*
  soil erosion, **9:**1155
indoor pollution
  hazardous household waste, **5:**626
  health effects, **5:**631
  nitrogen cycle, **1:**21–22, 23
  Pollution Probe, **8:**1005
  radon, **8:**1044, 1046, 1047
  sick building syndrome, **9:**1136–1137
  tobacco smoke, **1:**18–19, **9:**1222–1223
  unvented combustion risk, **1:**21
industrial ecology, **6:696–699,** *719*
  acid rain reduction, **1:***4*
  air pollution, **1:**18–19, *86,* 90–91
  alternative dispute resolution, **1:**30–31
  automotive manufacture, **1:**90–91
  Bhopal toxic gas-leak accident, **1:**108–109
  carbon and, **2:**155
  CFCs and, **2:**168, 169
  chemical industry, **2:**172–175
  clean coal technology, **2:**198–199
  climate change, **2:**188
  ecological literacy, **4:**435
  economic incentive, **3:**332, 337–339

emission standards, **2:**178–179, 197–198
green marketing, **5:**610–611
industrial metabolism, **6:**706–707
international projects, **6:**718–721
Lake Baikal, **1:**101
life cycle analysis, **6:**697, *699,* 770–771
poor communities, **2:**204–205
water pollution controls, **2:**180
industrialization, **6:700–705**
  of agriculture, **6:**704
  air pollution from, **1:**18–19, *86,* 90–91
  canal transportation, **2:**145
  energy resources, **2:***199*
  fossil fuel and, **5:**558
  geology and, **5:**582
  housing and, **6:**643, 644–645
  megacity development, **6:**808
  nitrogen fixation, **7:**881–882, 888
  noise pollution, **7:**890, 891
  occupational safety, **7:**899
  oxygen uses, **7:**930–931
  smog, **9:**1140, 1141
  technological changes, **9:**1192–1193
  technology-forcing regulation, **9:**1194–1195
  urbanization and, **10:**1306–1307
  utopianism as reaction to, **10:**1317
  water use, **10:**1354
  zoning, **10:**1399
industrial metabolism, **6:706–707**
  toxic metals, **9:**1224
Industrial Revolution. *See* industrialization
industrial waste. *See* hazardous waste, industrial
Indus Valley, **6:**751, **9:**1199, **10:**1330
infant mortality rate, **8:**1032
influenza epidemic, **10:**1321
information technology, **6:708–709**
infralittoral zone, **9:**1129
infrared radiation, **1:**84
infrared sensors, **8:**1079
injectable contraceptive, **2:**222–223
Inland Waterways Commission, **8:**1111
Inner Harbor Project (Baltimore). *See* Harborplace
inorganic compounds, **2:**153, 173
insecticide, **6:710–711,** 717
  synergistic effect, **9:**1187
insects and related arthropods, **6:712–717**
  chemical control of, 6.704, 6.710-711, 6.717. *See also* insecticide; pesticide
  control methods, **7:**964–965
  disease vectors, **3:**284–289, **6:**786, 787, **7:**960–961
  food irradiation against, **6:**725–727
  keystone mutualists, **6:**740
  parasitic, **7:**945, 947
  social insects, **9:**1144–1147, 1204–1205
  symbiotic microorganisms, **9:**1184–1185
  termites, **9:**1204–1205
insulation, **9:**1209
integrated pest management, **7:**961, 965
Integrated Solid Waste Management, **10:**1342–1343
Intelligent Vehicle/Highway Systems, **8:**1105, **9:**1241, 1244

Intercoastal Waterway, **4:***425*
interferon, **2:**193
intergenerational equity, **2:**217, **3:**339, **9:**1176
Intergovernmental Conference on Environmental Education (Tbilisi; 1977), **4:**418
Interior Department, U.S., **4:**541, 542, **5:**587, **7:**891, **8:**1021, **9:**1149, **10:**1313
Intermodal Surface Transportation Efficiency Act, **8:**1102, **9:**1245
Intermountain Bunchgrass, **8:**1053
internal combustion engine, **6:**702
Internal Revenue Service, **5:**667
international agreements. *See* environmental law, international; United Nations conference
International Atomic Energy Agency, **6:**725, **7:**895
International Bank for Reconstruction and Development, **10:**1294, 1295, 1388
International Biological Program, **10:**1283
International Commission on Radiological Protection, **8:**1045
International Convention for the Regulation of Whaling, **10:**1380–1381
International Convention on the Prevention of Pollution from Ships, **4:**428
International Council of Scientific Unions, **6:**718, **10:**1299
International Court of Justice, **10:**1299–1301
International Dark Sky Association, **6:**773
International Decade for Natural Disaster Reduction, **7:**857
International Development Association, **10:**1388
International Earth Day, **6:**797
International Environmental Education Programme, **4:**418
International environmentalism. *See* environmental law, international; United Nations conference; United Nations program
international environmentalism. *See* environmental law, international; United Nations conference; United Nations program
International Federation of Environmental Journalists, **6:**738
International Finance Corporation, **10:**1388
International Geophysical Year, **1:**45
International Geosphere-Biosphere Programme, **6:718–721**
International Global Atmospheric Chemistry Project, **6:**718, 721
International Group of Funding Agencies for Global Change Research, **6:**720, 721
International Human Dimensions Programme on Global Environmental Change, **6:**721
International Institute for Environment and Development, **10:**1328
International Joint Commission, **4:**427, 453
International Organization for Standardization, **6:**771
International Register of Potentially Toxic Chemicals, **10:**1297

Page numbers in **boldface type** indicate full articles on a subject. Page numbers in *italic type* indicate illustrations or other graphics. All numbers preceding colons indicate volume numbers.

Page numbers in **boldface type** indicate full articles on a subject. Page numbers in *italic type* indicate illustrations or other graphics. All numbers preceding colons indicate volume numbers.

Page numbers in **boldface type** indicate full articles on a subject. Page numbers in *italic type* indicate illustrations or other graphics. All numbers preceding colons indicate volume numbers.

Kirtland warbler nesting, **3:**324
mineral production, **7:**841
sport hunters, **5:**675
Michigan, Lake, **5:**601, 606
microbiology, **7:**956–957
microprocessors, **3:**386, 389
microwave, **6:820–821**
Middle East
archeology, **1:**62, 63
demographic transition, **8:**1010–1011
petroleum, **8:**980
See also specific countries and regions
Middleton, Paulette, *as contributor,*
9:1140–1143
Midgley, Mary, **1:**40
midocean drilling, **7:**907
midocean ridges, **5:**585
migrant labor, **6:822–823**, 824, 825,
826–827, **7:**902
migration, **6:824–827**
bird refuges, **7:**851, 852
demography, **2:**256
environmental equity and, **4:**461–462
exotic species and, **4:**479
forced by hydropower development,
5:685, **9:**1219
glaciers and, **5:**593–595
grasslands and, **5:**599–600
habitat, **7:**868, 869
lake, **6:**747
protected wildlife, **4:**426, 429
refugees, **6:**825, *826*
tundra, **10:**1282
worker. *See* migrant labor
Migratory Bird Conservation Act, **7:**852,
8:1021
Migratory Bird Hunting Stamp Act, **7:**852
Migratory Bird Treaty Act, **2:**137, **7:**852,
8:1021
Milford (Pa.), **8:**991
military activity. *See* war and military activity;
specific wars
Mill, John Stuart, **5:**649, **6:**791
Miller, Stanley, **7:**861
Milner, Richard B., *as contributor,* **6:**740–743
Milwaukee (Wisc.), **9:**1245
mimicry, **10:**1323
Minamata disease, **6:**815, **9:**1225
mind-body medicine, **6:**806–807
mineral cycle, **6:828–833**
mineralization, definition of, **7:**885
Mineral King Valley, **4:**423
minerals, **3:**278, 299, 327–328
public lands, **2:**140
See also mineral cycle; mining industry
Minerals Management Service, U.S., **5:**587
Mine Safety and Health Act, **7:**902
Mine Safety and Health Administration,
1:22, **7:**899, 902
Minick, Jill, *as contributor,* **8:**1000–1003
mining industry, **1:**98, **5:**597, **7:838–841**, **9:**1167
arid lands, **2:**265
bacteria and, **1:**98
coal, **2:***153,* 194, 195–197, *198,* 213
health and safety, **7:**899, 902
ocean resources, **7:***910*
radon risk, **1:**21–22
strip mining, **2:***153,* 213
tundra, **10:**1283

mini-trials, **1:**27
Minneapolis (Minn.), **7:**949, **10:**1309
Minnesota, **3:***364,* **8:**1066
Boundary Waters Canoe Area, **3:**321–323
major flood, **4:**510
mineral production, **7:**841
Sherburne National Wildlife Refuge,
8:*1021*
tree-planting buffer zone, **10:**1368
Min River, **9:**1200
missiles, long-range, **7:**913
"missing links," **4:**477
Mississippi, TVA mandate in, **9:**1202
Mississippi River, **1:**74, 75, **2:**242–244
Gulf of Mexico discharge, **5:**242–244
major floods, **4:**510
Mississippi River Delta, **2:***244,* **8:**1078, *1079*
Missoula, Lake, **5:**596
Missouri
cattle shipment centers, **8:**1048
major flood, **4:**510
mineral production, **7:**841
Missouri River, **1:**74–75
floods, **4:**510, *515*
Mitchell, Robert Cameron, *as contributor,*
4:442–447
mitochondria, **3:**292
Moby-Dick (Melville), **6:**775
Moet, Abdelsamie, *as contributor,*
8:1000–1003
Mohenjo Daro (Indus Valley), **10:**1330
Mojave Desert, **8:**1053, **10:**1279
solar mirrors, **3:***392*
molds, **5:**570–571
mole, definition of, **7:**888, 934
molecule, **1:**113
carbon, **2:**154
CFC, **2:**169
hydrocarbon, **7:**972
nitrogen, **7:***881*
oxygen, **7:**932, 934, 936, 937, 938
toxicology, **9:**1231
Moles, Mice, and Lemmings (Elton), **3:**361
Molina, Mario, **7:**940
Moll, Gary, *as contributor,* **4:**536–539
Monaco, GDP per capita, **8:**1027
monoculture, **9:**1254
monomer, **8:**1001, 1002
Monongahela National Forest, **4:**530
Monsanto company, **2:**173
Montana
buffalo restoration, **3:**373
coal reserves, **2:**197
Glacier National Park, **5:***596*
National Bison Range, **7:**851
Pleistocene Era flood, **5:**596
ranching, **8:**1048
rangeland, **8:**1049, *1050*
Monterey Bay Aquarium, **9:***1129*
Montezuma Castle monument, **7:**953
Monticello (Va.), **6:**758, 775
Montreal Protocol, **4:**428, **10:**1297
Moody, David W., *as contributor,* **2:**236–237,
5:606–607
Moon, **7:**926, *927*
More, Thomas, **10:**1316, 1317
Morgan, Thomas Hunt, **5:**577
Morgenthau, Henry, **2:**217
Morocco, **5:**620

Morse, Samuel, **6:**708
mortality rate. *See* death rate
Morton, James P., **8:**1065
Mortuary Temple of Queen Hatshepsut
(Luxor, Egypt), **6:***756*
Moses, Robert, **7:**749
mosquito, **3:**285–286, **6:**710, 717, 786, 787,
7:946, 960, 961, **10:**1321
Mother Earth, **1:**42
Motor Vehicle Air Control Act, **2:**179
Motor Vehicle Information and Cost
Savings Act, **5:**566
motor vehicles. *See* automobile; bus
Mount Brewer (Kings Canyon, Calif.), **4:***447*
Mount Laurel (N.J.) zoning, **10:**1401
Mount Olympus, **8:**1110
Mount Pinatubo volcano, **2:**185, **5:***585*
Mount Pinchot, **8:**991
mouse, **4:**479
Moveable Museum program, 7, *848,* **7:***849*
Mowat, Farley, **6:**780
Mozambique, **6:**825
Muhammad, Prophet, **8:**1070
Muir, John, **4:**422, 545, **6:**759, 776,
7:842–843, **8:**1110
and conservation movement, **2:***212,* 214,
4:437, 438, **7:**843
on equity of all creatures, **4:**461
legacy of writings, **6:**744, **7:**843
Pinchot controversy with, **7:**843, **8:**991
as preservationist, **7:**843, **8:**991, 1016,
1017, 1018
Sierra Club founding, **7:**843
and wilderness experience, **10:**1383
Yosemite National Park creation,
7:842–843, 953
Muir Woods National Monument, **8:**991,
1110
mule, **1:**34
Muller, Paul, **6:**710
Multilateral Investment Guarantee Agency,
10:1388
Multiple Use–Sustained Yield Act, **4:**540,
542, 543
multistage flash process (MSF), **2:**259
Mumford, Lewis, **6:**701, 777, **7:844–845**,
10:1331
Munich (Germany), **3:**307
Murchison, Roderick, **5:**583, **6:**784
Murdock, Barbara Scott, *as contributor,*
6:766–767, 814–815
Murmansk (Russia), **10:***1289*
Muscle Shoals (Ala.), **9:**1202, 1203
museums and exhibitions, **7.846-849.**
See also specific museums
mushroom, **5:**570, 571
Muskie, Edmund, **4:***421,* 424, 448
Muslims. *See* Islam
mutation, **3:**406, **5:**578, 630, **9:**1230–1231
as adaptation, **1:**7
mutualism, **3:**331
tropical rain forest, **10:**1268–1269
mycotoxins, **3:**280
Myers, Norman, *as contributor,* **2:**164–165
Mykonos (Greece), **3:***397*

Nabhan, Gary Paul, **6:**781
Nader, Ralph, **3:**306
Naess, Arne, **2:**248

Nairobi (Kenya), **10:**1296
Namibia, Walvis Bay, **5:***620*
nanoplankton, **1:**84
naphthalene, **7:**972
Napoleon, emperor of France, **8:**1100
Nasser, Lake, **1:**81
Nathans, Daniel, **5:**653
National Academy of Science, **3:**283, **7:**900
National Advisory Environmental Health
  Committee, **7:**874
National Aeronautics and Space
  Administration, **5:**586, **7:***938, 939,*
  10:1314
National Ambient Air Quality Standards,
  2:178, 278
National Audubon Society, **1:**89, **4:**423, 436,
  442, *445,* 469, **6:**745, **7:**851, 969
National Biological Information
  Infrastructure, **5:**587
National Bison Range, **7:**851
National Center for Atmospheric Research,
  8:1106–1107
National Council on Radiation Protection
  and Measurements, **8:**1045
National Earthquake Hazards Reduction
  Program, **5:**586
National Environmental Education Act,
  4:416
National Environmental Education
  Advocacy Project, **4:**418
National Environmental Policy Act, **4:**424,
  433, 450, **5:**667, **6:**753, **8:**1034,
  10:1311
  aesthetic values influence, **4:**465
  air and water pollution requirements,
    1:75, **2:**179, 181
  Clean Fuel Fleet program, **5:**557
  environmental assessment process,
    4:412–415, 542–543
  as first major environmental statute,
    4:424
  highway projects, **8:**1104
  nuclear power plant licensing impact,
    7:896, 897
  technology-forcing regulation, **9:**1194
National Forest Management Act, **4:**465,
  540
National Highway System, **8:**1102
National Historic Landmarks, **2:**140
National Institute for Occupational Safety
  and Health, **7:**874, 900, 901, 903, 904,
  8:1136
National Institutes of Health, **2:**133, **5:**654,
  7:900, **8:**1030
National Interagency Fire Center, **2:**141
nationalism, **6:**826
National Monuments, **2:**141
National Monuments Act, **8:**1110
National Occupational Exposure Survey,
  7:900
National Oceanic and Atmospheric
  Administration, **4:**500, **10:**131, 1311
national parks. *See* park, U.S. national
National Parks and Conservation
  Association, **4:**442, **6:**773

National Park Service, **4:**436, 468, **5:**587,
  6:753, 759, 761, 773, **7:**952–955,
  10:1313
  creation and guiding principle of, **7:**952
  See also park, U.S. national
National Pollutant Discharge Elimination
  System, **10:**1369
National Register of Historic Places, **2:**140
National Religious Partnership for the
  Environment, **8:**1066–1067
National Renewable Energy Laboratory,
  3:397
National Resource Conservation Service,
  6:764
National Resources Defense Council, **3:**365,
  4:443, **8:**1034–1035, 1037
National Resources Planning Board, **10:**1387
National Scenic Byways program, **8:**1102
National Science Foundation, **5:**654, **10:**1314
National System of Interstate Highways,
  8:1102
National Toxics Campaign, **4:**447
National Trails system Act, **4:**465
National Wildlife Federation, **4:**442
National Wildlife Refuge Administration
  Act, **8:**1021
National Wildlife Refuge System, **7:**850–853,
  866, 953, **8:**1021, 1110
National Wildlife Refuge System
  Improvement Act, **8:**1021
Native Americans
  animism, **1:**43
  Colorado River holy sites, **9:**1201
  effects of seal-hunting ban on, **1:**53
  Eskimo and Inuit, **1:**53, **5:**594, **8:**1062,
    10:1282, 1283
  land rights, **6:**750–751
  religion, **8:**1062
  Toltec burial grounds, **1:***53*
  totem pole, **1:***49*
  Trans-Alaska Pipeline effects on, **9:**1234
natural balance. *See* ecological stability
natural disaster, **7:**854–857, 910, 913
  avalanche, **5:**595–596
  from El Niño, **3:**356–357
  See also flood; hurricane; volcanic eruption
natural gas, **5:**564, **7:**858–863
  as automotive alternative fuel, **1:**91–92
  carbon in, **2:**152, 153, 157
  cogeneration techniques, **2:**202
  as energy source, **3:**377, 378
  as fossil fuel, **5:**558–559
  industrialization and, **6:**705
  liquefication, **5:**556, 560–561, **7:**858,
    859–860, 861, 863
  methane in, **5:**558, **7:**858, 859, 862
  petrochemicals, **2:**173
  petroleum and, **7:**972, 974, **8:**983
  synthetic, **5:**562
natural habitat, **7:**864–869
  biodiversity and, **1:**114
  biotic adaptability, **2:**131
  of birds, **2:**137
  coral reefs, **2:**224
  dam building effects, **2:**233–235

  ecological change and, **3:**324–325
  Elton's animal studies of, **3:**360, 361
  endangerment from loss of, **3:**341, 371,
    372, 373
  grassland, **7:**864, 868, 869
  human impact on, **1:**114
  keystone species, **6:**740–743
  niche, **7:**875–877
  oil spill hazard, **7:**915
  preservation, **8:**1037
  river and stream, **8:**1098–1099
  tropical rain forest, **10:**1264
  water availability, **10:**1359
  waterfront development effects, **10:***1360*
  wetland protection, **10:**1368, 1372–1373,
    1376
  wildlife refuges, **7:**850–853
Natural History Museum (Oxford), **10:***1286*
Natural Resources Conservation Service,
  1:107, **6:**762–763, **10:**1311
Natural Resources Defense Council, **4:**423,
  437, **8:**1034, 1037
natural selection, **3:**331, 406, **4:**473, 474,
  10:1333
nature
  balance in, **3:**320
  Emerson's writings on, **3:**363, **9:**1214
  human rights vs., **5:**665, 666
  landscape painting, **4:**463–464
  writing, **6:**774–776
  See also ecological stability; wilderness
    experience
Nature (Emerson), **3:**363, **9:**1214
Nature Conservancy, **4:**446, **9:**1212
Nature Pleads Not Guilty (Garcia), **8:**1107
nature vs. nurture controversy, **7:**870–873
naturopathy, **6:**805
Navajo, **6:**750–751
Nebraska
  cattle shipment centers, **8:**1048
  irrigated cropland, **5:**603
  major flood, **4:**510
  pioneers, **6:***750*
  primitive road, **8:***1101*
  range management, **8:**1054
  remote sensing information, **8:**1077
  sod house, **5:***598*
nebula, **3:***355*
negotiated rulemaking ("reg-neg"), **1:**27
negotiation, **1:**27
Nelson, Gaylord, **3:**306
Nelson, Norton, **7:**874
Nelson, Robert H., *as contributor,* **6:**754–755
Nepal, **9:**1153
  cotton field, **8:***996*
  fertility rates, **8:**1010
Netherlands
  canals, **2:***144*
  chemical plant explosion, **2:***173*
  North Sea flood, **4:**510
  tidal power, **3:**395
  waste disposal, **10:**1336
Netiv-Hagdud site (Jordan Valley), **8:**996
Neuse River, **8:**1035
neutral evaluation, **1:**27
neutron, **7:**893
Nevada
  Great Basin Desert, **2:***260*
  mineral production, **7:**841

Page numbers in **boldface type** indicate full articles on a subject. Page numbers in *italic type* indicate
illustrations or other graphics. All numbers preceding colons indicate volume numbers.

ranching, **8:**1048
rangeland, **8:**1049, 1053
Reno, **6:***763*
New Deal, **6:**759, **8:**991
   Civilian Conservation Corps, **4:**541,
      8:991, **9:**1149
   conservation movement, **2:**217
   soil conservation, **9:**1147, *1148*, 1149
   suburban housing, **9:**1162, 1163
   See also Dust Bowl
Newell, Frederick, **8:**1111
New England
   colonial-era tree protection, **4:**544
   landscape esthetics, **4:**464
   Light Pollution Advisory Group, **6:**773
   radon levels, **8:**1047
   transcendentalism, **3:**362, 363, **9:**1236
   whaling industry, **10:**1379–1380
   See also specific states
Newfoundland, **10:**1379, 1380
New Guinea, grassland, **1:**123
New Hampshire, Hubbard Brook Ecosystem
   Study, **10:**1370, 1371
New Jersey
   garden suburbs, **5:**609, **10:**1307, 1308
   Mount Laurel zoning, **10:**1401
   radon levels, **8:**1047
   rapid transit system, **9:**1242
   urban problems, **10:**1309
Newlands Reclamation Act, **2:**181, **8:**1110
New Look at Life on Earth, A (Lovelock),
   5:572, *573*
New Mexico
   energy-efficient building, **3:***401*
   Gila Wilderness Area, **6:**768–769
   Los Alamos nuclear site, **7:**893
   rangeland, **8:**1049, 1053
New Orleans (La.), historic area protection,
   4:464
Newport News (Va.), **7:***858–859*
new towns, **5:**608–609, **9:**1163, **10:**1302,
   1303, 1304
New York City, **3:***328*, **6:***809*
   American Museum of Natural History,
      1:*113*, **6:***796*, **7:***847*, 848, 849
   Battery Park City, **10:**1302, 1361
   Brooklyn, **7:**917, 949, **9:***1135*
   Central Park, **4:**464, **6:**758, 761, **7:**916,
      917, 949, 953
   commuter Clean Air requirements, **9:**1245
   Earth Day rally, **3:**306–307
   first buses, **8:**1101
   labor movement, **6:**744
   Lever House, **1:***66*
   parks and recreation design, **7:**949
   parkway system, **8:**1103
   Regional Plan of 1929, **10:**1308
   Roosevelt Island, **10:**1302
   subway system, **9:**1242, 1246
   UN headquarters, **10:***1298, 1299, 1301*
   urban forests, **4:***536*, 549
   waste disposal, **10:**1332, 1333
   zoning ordinances, **10:**1308, 1398, 1399
   Zoological Society, **6:**777
New York Fish Commission, **8:**1110
New York Sportsmen's Club, **2:**216
New York State
   Adirondack Park, **6:**775
   ecolabeling, **3:**314

energy efficiency program, **3:**399
Erie Canal, **1:**74
landscape esthetics, **4:**464
Levittown, **4:***507*, **9:**1162
Long Island Pine Barrens, **1:**31
Love Canal, **2:**203, **4:**439, 447, **5:**628,
   6:782–783, **9:**1168
   major flood, **4:**510
   Niagara Falls, **6:**782, 783, **7:**917
   Olmsted projects, **7:**916, 917
   Oneida community, **10:**1316
   radon levels, **8:**1047
   remote sensing information, **8:**1077
   Shoreham Nuclear Plant, **8:**897
   sport hunters, **5:**675
   Storm King Mountain, **1:***26*, 28, **4:**422,
      423
New York State Lung Association, **6:**744
New York Zoological Society, **3:**373, **6:**777
New Zealand
   algal bloom, **8:***995*
   ranches, **8:**1048
   temperate rain forest, **9:**1196
   tussock, **1:**119
Niagara Falls (N.Y.), **6:**782, 783, **7:**917
Nicaragua, fertility rate, **8:**1010
niche, **7:**875–877
nicotine, **9:**1222, 1223
Nierenberg, William, *as contributor*,
   7:904–913
Nietzsche, Friedrich, **5:**649
Nigeria, percent of underweight children,
   5:673
Nile basin, **5:**600, **6:**751, **9:**1199
Nile River, **1:**80, *81*, **2:**144, **6:**728, **9:**1199,
   10:1330
   flooding, **4:**509, 510, **5:**638–639
NIMBY, **7:**878–879
   environmental assessment and, **4:**415
   environmental law and, **4:**424, 425
   grassroots ecologism, **4:**438, 439
Nirenberg, Marshall, **5:**653
nitrate/nitrite, **7:**880–885, 889
   level in water, **10:**1367
nitrogen, **2:**197, 228, **7:**859, **880–882**
   as basic element, **3:**354–355
   carbon and, **2:**153, 154, 155
   compost ratio, **2:**210, **10:**1339
   distribution of, **7:***884*
   emissions, **9:**1249, 1250
   fertilizer, **4:**486, 487, **5:**620, **7:**881, *885*,
      8:1108, **9:**1192
   liquid, 556
   oxygen and, **7:**930
   percent in earth's atmosphere, **5:**573
   water quality, **10:**1367
nitrogen cycle, **3:**328, 343, **7:**880–881, **883–889**
   acid rain formation, **1:**2, 3–4, 87
   algae and, **1:**24–25
   aquaculture effects, **1:**57
   in atmosphere, **1:**82, 84, 87
   bacteria and, **1:**96–97
   dead zone, **2:**242, 243, 244, 245
   eutrophication, **3:**291, **4:**466
   fungus, **5:**570
   indoor air pollution, **1:**21–22, 23
   industrial ecology, **6:**696
   legume-rhizobia symbiosis, **9:**1182–1184
   wetland action, **10:**1374

nitrogen dioxide, **2:**197, **7:**880, 929, 931
nitrogen fixation, **7:**886–888
nitrogen oxide, **2:**159, 169, **4:**452
   Clean Air Act, **2:**178
   coal and, **2:**199
   combustion-produced, **10:**1339, 1340
   emissions control, **1:**92
   nitrogen cycle, **7:**880, 885, 888, 889
   oxygen and, **7:**929, 930
   petroleum emissions, **8:**982
Nixon, Richard, **4:**424, 448, 449, 543
Noah's Ark, **8:***1072*, 1073
Nobel, Claes, **8:**1065
Noguchi, Isamu, **7:**950
noise pollution, **7:**890–891, 900
   hearing loss, **5:**633
Noise Pollution Control Act, **4:**424, **7:**890
nomads, **3:**270
nongovernmental organizations, **2:**247
non-Hodgkin's lymphoma, **5:**636–637
nonrenewable resource. *See* economics
   of renewable and nonrenewable
   resources; fuel, fossil
Nordhaus, William D., **3:**346
   as contributor, **8:**1022–1025
Norfolk (Va.), waterfront renovation, **10:**1303
North Africa
   demographic transition, **8:**1010–1011
   domesticated animals, **1:**33, 34
   windmills, **9:**1190
   See also specific countries
North American Association for
   Environmental Education, **4:**418
North American Conservation Conference,
   8:991
North Atlantic Marine Mammal
   Commission, **10:**1381
Northbrook (Ill.), **6:**726
North Carolina
   Biltmore estate, **7:**917
   flood, **4:***511*
   Great Smokey Mountain National Park,
      9:1142
   Hurricane Floyd, **8:**1035
   plant accident, **7:**902
   TVA mandate in, **9:**1202
   Venus flytraps, **3:**373, **4:***420*
   Wilmington waterfront development,
      10:1361
North Dakota, major flood, **4:**510
North Denver (Colo.), **9:***1227*
Northeast corridor, **9:**1243
Northeast Passage, **1:**72
northern spotted owl, **4:**533, 546
North Sea
   flood, **4:**510
   oil fields, **8:***980*
   oil spills, **6:**705
North Slope (Alaska), **9:**1234, 1235, **10:**1280,
   1283
Northwest Ordinance of 1787, **6:**751
Northwest Passage, **10:**1282
Northwest Territory (Canada), **10:**279
Norway
   GDP per capita, **8:**1027
   glaciers, **5:**594, 595
   labor movement, **6:**745
   temperate rain forest, **9:**1196
   whaling industry, **10:**1379, 1380, 1381

notation, scientific, 7:935
Notre-Dame-du-Haut (France), 1:*67*
Novak, John T., *as contributor*, 2:258, 3:298–299
Nova Scotia, 3:307, 10:1352
  tidal power, 3:395
nuclear power, 3:377, *381*, 7:**892–897**
  Arctic icebreakers, 1:72
  carbon as catalyst, 2:155
  Chernobyl disaster, 1:73, 2:176–177, 205, 6:705, 7:896, 8:1046, 10:1326
  electric utility industry, 3:349, 350–351, 352, *353*, 389
  fission energy, 3:379
  fusion energy, 2:155, 3:380
  as military target, 10:1325–1326
  public health safeguards, 10:1387
  radioactivity, 8:1042–1043
  thermal pollution, 9:*1211*
  Three Mile Island accident, 2:205, 6:705, 7:895, 897, 9:1220–1221
Nuclear Regulatory Commission, 1:28, 3:349, 7:897, 8:1045, 9:1220, 10:1313
nuclear weapons, 7:892–893, 895, 913
nucleic acids, 7:881
nuisance disputes, 1:30–31
nuisance law, 4:433
nutrient cycles. *See* mineral cycle
nutrition. *See* diet and health
nylon, 2:174

Oaxaca (Mexico), 8:996
obesity, 3:277–278
occupational safety and health, 5:631, 633–636, 7:**898–903**
  air pollutants, 1:19, 22
  asbestos risks, 1:77
  carbon monoxide poisoning, 2:161
  chemical industry, 2:172, 174
  factory safety laws, 7:898–899
  Nelson's work, 7:874
  petroleum industry, 8:982–983
  radon risk, 8:1044, 1046
  sick building syndrome, 9:1136–1137
  workers' compensation, 2:179
Occupational Safety and Health Act, 6:744, 745, 7:899, 901
Occupational Safety and Health Administration, 1:22, 77, 2:161, 174, 7:703, 899, 900, 901–902, 903, 10:1314
  Environmental Network, 6:745
  establishment of, 7:899
ocean
  carbon cycle, 2:158
  currents, 7:911
  Earth Observing System data, 3:311
  ecological pyramid, 3:317, 318
  effects on atmosphere, 2:186
  El Niño effects, 3:356–357
  as energy source, 3:395, 8:1080, *1081*, 1084, 1085
  fishing industry, 4:498–501
  greenhouse effect, 5:613, 616, 617
  international measures, 4:428–429

inter-ocean canal effects, 2:145
layers, 7:911
oil deposits, 7:910, 974
oil spill, 7:914, 915
photosynthesis, 8:988
seawater farms, 7:924–925
shoreline, 9:1132
tidal energy, 3:395, 8:1080, *1081*, 1085
UN Law of the Sea Conference, 10:1291–1292
water cycle, 10:1353, 1355, 1356, 1357
zones, 9:1128–1131
See also aquatic life; oceanography
Ocean Dumping Convention, 4:427, 428
Oceanic Society, 4:443
oceanography, 7:**904–913**
  Carson's contribution, 2:166
  Cousteau's contribution, 2:231
  geologic mapping, 5:585, 586
  international projects, 6:718
  sulfur cycle, 6:833
octane, 7:972
Odiello Solar Research Center, 8:*1082*
Odum, Eugene P., 3:347, 7:875
  as contributor, 3:326–331
off-gassing, 1:18
Office of Economic Ornithology, 8:1021
Office of Environmental Education, 4:416, 417
Office of Management and Budget, 10:1311
Office of Science and Technology Policy, 10:1311
Office of Surface Mining, U.S., 5:587
Office of the U.S. Trade Representative, 10:1311
office park, 9:1163
Ogallala Aquifer, 1:59, 4:502
Ogino, Kyusaku, 2:222
Oglala Sioux, 1:43
Ohio
  Euclidean zoning, 10:1399, 1400
  wind turbine, 3:*396*
Ohio River, 1:74
oil. *See* petroleum; petroleum industry
oil crises (1970s), 1:69, 7:921, 8:981
Oil Pollution Act, 6:736
oil shale, 5:562–563
oil spill, 4:*436*, 7:**914–915**, 975
  DDT synergistic toxic effect, 9:1187
  Exxon *Valdez*, 4:444, 6:736, 737, 9:1235
  hazardous household waste, 5:624–625
  industrial, 6:705
  journalist reports, 6:736, *737*
  recreational activity, 8:1058
  risk reduction, 8:983
  war-induced, 10:1325
Okeechobee, Lake, 4:468, 469, *480*
Okefenokee National Wildlife Refuge, 8:*1058*
Oklahoma
  buffalo restoration, 3:373
  natural gas reserves, 7:858
  Wichita Mountains Preserve, 3:373, 7:851
Old Crow (Canada), 10:1283
Olduvai Gorge (Tanzania), 4:*474*

oligotrophic lake, 4:466, 467
Olkhon Island, 1:101
Olmsted, Frederick Law, 4:464, *539*, 6:758, 759, 7:**916–917**, 948, 949, 953
  belief in parks as social change agent, 7:948, 949
Olson, Sigurd, 6:777
Olympic National Forest, 4:*545*
Omenn, Gilbert S., *as contributor*, 2:135, 8:1028–1033
Oneida (N.Y.) community, 10:1316
1000 Friends of Oregon, 8:1035
Ong Tsui, Amy, *as contributor*, 4:482–485
Only One Earth: The Care and Maintenance of A Small Planet (Dubos), 3:304
Ontario (Canada), 2:145, 5:606, *644*, 8:1004
Ontario, Lake, 5:606
On the Origin of Species (Darwin), 2:239, 4:472, 473, 5:577, 583
oocyte, 2:191, 192
OPEC, 1:69, 3:352, 7:**918–923**
open-cast (pit) mining, 7:839, 840
open hearth process, 6:702
Operation Desert Bloom, 2:*262*, 7:**924–925**
Oppenheimer, Robert, 7:893
Oracle (Ariz.), 3:*345*
oral contraceptive, 4:484
Orange County (Calif.), 5:625
orangutan, 1:49
Oregon, 8:1035
  architecture, 1:68
  as environmental reform–land-use model, 6:794–795
  natural storage, 7:*863*
  rangeland, 8:1049, 1053
  See also Portland
Oregon Trail, 8:1101
organic farming, 8:1108–1109, 9:*1170*, 1172, 1173–1174
organic fertilizers, 4:486
organic Rankine turbines, 2:202
Organization for Economic Cooperation and Development, 8:1038, 1039
Organization of Petroleum Exporting Countries. *See* OPEC
Origin of Species (Darwin). *See* On the Origin of Species
Origin of the Alps, The (Suess), 9:1165
Orissa (India), 5:568, 7:857
Ornish, Dean, 6:807
Ornithological Biography (Audubon), 6:775
Orwell, George, 10:1317
Osaka (Japan), 10:1303
Osborn, Henry Fairfield, 6:777
O'Shaughnessy Dam, 8:1018
osteoporosis, 3:278
Ottawa (Canada), 8:1004
Our Common Future (1987 report), 2:217
Our Plundered Planet (Osborn), 2:216
outdoor recreation. *See* recreation, outdoor
outer space, 7:**926–927**
  cosmic radiation, 8:1043–1044
  remote sensing, 8:1076–1079
ovarian cancer, 2:162
Overland Trail, 8:1101
ovulation, 5:662–663
Owen, Wilfred, *as contributor*, 10:1302–1305
oxidation, 7:936
oxides, 7:935

Page numbers in **boldface type** indicate full articles on a subject. Page numbers in *italic type* indicate illustrations or other graphics. All numbers preceding colons indicate volume numbers.

Page numbers in **boldface type** indicate full articles on a subject. Page numbers in *italic type* indicate illustrations or other graphics. All numbers preceding colons indicate volume numbers.

Pollio, Marcus Vitruvius, **10**:1357
pollution. *See* air pollution; indoor pollution; light pollution; noise pollution; water pollution
Pollution in Paradise (film), **6**:*794*, 795
Pollution Probe, **1**:*19*, **8**:**1004–1005**
pollution trading, **4**:425, **8**:**1006–1007**
  acid rain, **1**:2
  economic incentive, **3**:332–333, 336–338
  tradable permit, **9**:1232–1233
Polo, Marco, **9**:1134
polychlorinated biphenyl. *See* PCB
polyethylene, **8**:1001, 1002, 1003
polymerization, **8**:1001–1002
polymers, **8**:1000–1002
polyvinyl chloride, **2**:174, **8**:1001, 1002
Pompeii, **7**:845, **10**:*1334*
Ponce de Leon, F. Abel, *as contributor*, **2**:190–193
pond, **7**:865
Popper, Frank J., *as contributor*, **7**:878–879
population, **8**:**1008–1015**, 1041, 1054
  animal ecology, **3**:360–361
  as biosphere resources pressure, **2**:128
  carrying capacity, **2**:165, **3**:331
  demographic processes, **2**:256–257
  ecological changes in, **3**:323–325
  ecologists' study of, **3**:326
  endangerment, **3**:370–373
  energy consumption, **3**:*377*
  family planning, **2**:223, **4**:482–485
  food shortage, **5**:668, 669, 670, *673*
  growth rates, **9**:1177–1178, 1178–1179
  growth warnings, **6**:777, 778
  human reproduction, **5**:662, 663
  industrialization effects, **6**:700
  Malthusian theory, **6**:788–789
  megacity, **6**:808–809
  migration, **5**:685, **6**:824–827, **9**:1219
  projections by regions and countries, **8**:1015
  resettlement of, **5**:685, **9**:1219
  rural community, **8**:1112, 1115
  soil erosion rates and, **9**:1153
  sustainable agriculture and, **9**:1171–1172
  zero growth, **8**:1012–1013, 1033
Population and Development, UN Conference on, **8**:*1014*, **10**:*1291*, 1293
Portland (Ore.), **1**:68, **7**:950, **10**:1309
  Light Rail Transit system, **9**:*1247*
Port Sunlight (Liverpool suburb), **10**:1307
potassium, **8**:1042, 1043
  fertilizer, **4**:486, 487
Pou-Son-Tung, **1**:103
Poussin, Nicolas, **4**:463, **5**:657
poverty
  health and, **5**:632
  homelessness and, **5**:640–641
  hunger and, **5**:668, 669–670, 671, 672
  sustainable development and, **9**:1178, 1179
Powder River basin, **7**:840
Powell, Jerry, *as contributor*, **6**:792–793
Powell, John W., **5**:586
precipitation. *See* water cycle
predators, **2**:208, 209, 215–216, 226, **3**:316, 318, 330
pregnancy. *See* human reproduction
preservation vs. conservation, **8**:**1016–1021**
  Pinchot as conservationist, **8**:990–991
  public interest law, **8**:1036

Presidential Council on Environmental Quality, **4**:424
press. *See* journalism
Pribyat (Ukraine), **7**:896
price system, **8**:**1022–1025**, 1117
  as energy consumption curb, **9**:1180
  productivity, **8**:1026, 1027
  scarce goods allocation by, **9**:1126–1127
Priestley, Joseph, **7**:880, 928, 929
Priestly, Keith, **4**:*446*
primate ecology, **1**:49–51
Prime, Ranchor, *as contributor*, **8**:1068–1071
Prince Edward Island (Canada), **3**:*400*
Prince William Sound, **6**:736, **9**:1235
Principles of Geology (Lyell), **5**:583, **6**:784, 785
printing press, **6**:708
Pritchard, Paul C., *as contributor*, **7**:952–955
productivity, **8**:**1026–1027**, **9**:1154–1155, 1190
  labor, **2**:150
profundal zone, **9**:1130
Progress for a Small Planet (Dubos and Ward), **10**:1328
propane, **5**:556, 561
property rights, **4**:462, **6**:751, 754–755, 788–789
Prospect Park (Brooklyn, N.Y.), **7**:917, 949
Prosser, Norville S., *as contributor*, **4**:496–497
Protection Act (1894), **8**:1110
protein
  nitrogen content, **7**:881
  phosphorus synthesis of, **8**:984
  photosynthesis process, **8**:987
  sulfur and, **9**:1166
  viral, **10**:1321
Protestant Reformation, **8**:1075
protists (protoctists), **8**:993
Providence (R.I.), **7**:949
Prudhoe Bay, **9**:1234, 1235, **10**:1278, 1283
public health, **2**:177, 178–179, **5**:631, 632, **8**:**1028–1033**
  AIDS, **1**:16–17
  bioassay risk assessment, **1**:110–111
  demographic transition, **8**:1008–1011
  diet and nutrition, **3**:279–381
  drinking water safety, **10**:1386
  environmental, **7**:874
  environmental engineering, **10**:1386, 1387
  epidemiology, **4**:455–459
  family planning programs, **4**:482–485
  farm wastes, **8**:1035
  food irradiation benefits, **6**:727
  industrial toxic gas leak, **1**:108–109
  lead poisoning, **6**:766–767
  Love Canal, **6**:783
  malaria cases, **6**:787
  noise pollution, **7**:890–891
  occupational, **7**:874, 898–903
  PCBs and, **7**:958–959
  risk, **8**:1091–1085
  smoking hazards, **9**:1222–1223
  trace insecticide threats, **6**:711, 717
  tuberculosis, **10**:1276–1277
  urban forest benefits, **4**:536
  viral infection, **10**:1321
  waste disposal, **10**:1330–1335
  See also disease, vector-borne

Public Health Service, U.S., **2**:178, **7**:874, **8**:1030, **10**:1276, 1369
  agencies of, **8**:1030
  National Institute for Occupational Safety and Health, **7**:901
  nitrate levels, **7**:882
  noise exposure risks, **7**:890
public interest, **8**:**1034–1037**
  acid rain problem, **1**:4–5
  environmental law and, **4**:424–425
  environmental literacy as, **4**:434–435
  national forests and, **4**:540–543
  pricing system and, **8**:1024–1025
  risk, **8**:1099–1095
public lands, **6**:752–753, 763, 765, 776
  Bureau of Land Management, **2**:140–141
  conservation movement, **2**:213–214, 216
  historical misuses of, **2**:213
  See also forest, U.S. national; National Wildlife Refuge System; park, U.S. national
Public Utility Holding Company Act, **3**:352
Public Utility Regulatory Policies Act, **3**:350, 352, 401
pyramids, **10**:1397
pyrite, **6**:832

Qatar, **7**:923
quality of life, **3**:353, 389, **8**:**1038–1041**
  benefit-cost analysis, **1**:104–105
  consumption and, **2**:218
  deep ecology and, **2**:248–249
  noise pollution and, **7**:890–891
  outdoor recreation and, **8**:1057–1059
  overpopulation fears, **8**:*1012*
  rural advantages, **8**:1114–1115
  sustainable development, **9**:1176
Quammen, David, **6**:781
Quarantelli, E. L., *as contributor*, **7**:854–857
Queensland (Australia), **5**:604
Quesnay, François, **8**:1023
quilt, AIDS, **1**:*16–17*
quinine, **6**:799

Rabalais, Nancy, *as contributor*, **2**:242–245
rabbit, **1**:35
racial issues. *See* equity
Radburn (N.J.), **5**:609, **10**:1308
radiation. *See* irradiation, food; ultraviolet radiation; X ray
radiation, ionizing, **8**:**1042–1045**
  as cancer therapy, **2**:148
  Chernobyl accident, **2**:176, 177
  fissioned material, **7**:893
  food preservation, **6**:722–727
  greenhouse effect, **5**:612–617
  microwave, **6**:820–821
  nuclear debris, **10**:1326
  nuclear waste, **7**:895
  occupational safety standards, **7**:899
  radon risk, **8**:1046–1047
  Three Mile Island, **9**:1220
radiation sickness, **2**:177
radical environmentalism, **4**:*430*, 438, 440, 446
radioactivity, **8**:1042–1043
radium, **8**:1046
radon, **8**:1044, **1046–1047**
  lung cancer risk, **1**:21–22
  radiation risk, **8**:1044

Page numbers in **boldface type** indicate full articles on a subject. Page numbers in *italic type* indicate illustrations or other graphics. All numbers preceding colons indicate volume numbers.

Page numbers in **boldface type** indicate full articles on a subject. Page numbers in *italic type* indicate illustrations or other graphics. All numbers preceding colons indicate volume numbers.

Social Security Act, **7:**899
Society for Range Management, 1050, **8:**1050
Society of Environmental Journalists, **6:**738
sociobiology, **7:**873
soft water, **3:**274, 275
soil
  horizons, **6:**749–750
  lead contamination, **6:**766
  mineral cycle, **6:**831–832, 833
  moisture replenishment, **10:**1356
  nitrogen content, **7:**883–886
  tropical rain forest, **10:**1267–1268
  See also soil conservation; soil erosion
Soil and Water Conservation Society, **1:**107
soil conservation, **3:***365*, **9:1148–1151**,
    1153–1154, *1156*, 1157
  agricultural, **1:**8, 10, 13, 106–107
  agroforestry, **9:***1176*
  animal manure, **1:**33
  Bennett's work, **1:**106–107
  composting, **2:**210, 211
  conservation movement, **2:**213
  contamination liability resolution,
    1:29
  desertification and, **3:**273
  fertilizer and, **4:**486–487
  forest operations and, **4:**530
  Rodale's work, **8:**1108–1109
  sustainable agriculture, **9:**1172–1173
  TVA projects, **9:**1202
Soil Conservation Service, U.S., **1:**106–107,
    4:541, **9:**1149, **10:**1369
soil erosion, **6:**749, **9:1142–1157**, 1148, *1149*,
    1150, 1151, **1152–1157**
  from agriculture, **1:**8, 10, 52
  from Aswan High Dam, **1:**80, 81
  Clean Water Act reduction in, **2:**180
  from clear-cutting, **2:**182
  conservation movement, **2:**217
  coral reef effects, **2:**227–228
  from deforestation, **2:**255
  desertification and, **3:**271–272
  drought and, **3:**301
  grasslands, **5:**603
  water cycle, **10:**1359
  watershed, **10:**1370
  See also soil conservation
Soil Erosion Service, U.S., **9:**1149
soil microbe, **3:**304
solar cell, **3:***399*, **8:**1083, **9:1158–1159**
  electric voltage from, **5:**555–556, 557
solar energy, **3:**352, 353, *366*, 377
  assessment of, **3:**379
  collectors, **3:***392*, 393–394
  as renewable energy source, **8:**1080, 1081,
    1082–1083
  water cycle, **10:**1355, 1357
  Wright house heating design, **10:**1390
solar houses, **1:**67, 68, **3:***400*, **10:**1390
solar physics, **8:**1106
Solid Waste Disposal Act, **10:**1335, 1341
Somalia, **5:**671, 690, **6:**825, **8:**1011, 1051
Somers, David A., *as contributor,* **2:**133–135
Sonoran Desert, **8:**1053
sound pressure level, **7:**890
source-sink situation, **3:**327
South Africa, El Niño effects, **3:**357
South Asia, birthrate decline, **8:**1010

South Carolina
  Hilton Head Island, **4:**423
  shoreline settlement, **9:**1134
  Venus flytrap, **3:**373
  See also Charleston
South Coast Air Quality Management Plans,
    8:1105
South Dakota
  Grand River National Grassland, **5:***600*
  major flood, **4:**510
Southern California Air Quality
    Management District, **8:**1007
Southern Environmental Law Center, **8:**1035
South Georgia Island, **10:**1278
South Pole, **1:**44
Soviet Union (former)
  Aral Sea, **1:**60–61, **6:**705
  El Niño effects, **3:**357
  fishing industry, **4:**498
  natural gas reserves, **7:**858
  nuclear bombs, **7:**893
  petroleum resources, **8:**980
  planned community, **10:**1317
  See also Russia; Siberia; *other former
    republics*
Soylent Green (film), **10:**1317
space. *See* outer space; satellites
Space Needle (Seattle), **5:***670*
Spaceship Earth concept, **5:**572–573
Spain
  Barcelona project, **7:**950–951
  Basque region pasture land, **6:***762*
  Moorish garden design, **6:**757
  savanna grassland, **1:**119
spark-ignition engines, **2:**202
sparrow, **4:**479
species
  biodiversity, **1:**112, 113–114, **2:**217
  biome classification by, **1:***121*
  biota, **2:**130, 131
  birds, **2:**137
  carrying capacity, **2:**164–165
  classification by, **1:**96
  community, **2:**206–209
  coniferous forests, **4:**523–524
  Darwin's origins theory, **2:**239–240
  deciduous forest, **4:**531
  evolutionary, **4:**475–476
  exotic, **4:**478–481, 531, **10:**1362–1363
  Gaia hypothesis, **5:**573
  genetics, **5:**576, 577–579
  human ecology, **5:**650
  pesticide-resistant, **2:***134*
  water diversion effects, **2:**145, 233
  See also endangered species; extinction
Species Survival Commission, **6:**733
specific heat, water, **10:**1352
speech. *See* human communication
Spencer, Frank, *as contributor,* **10:**1286–1287
Spencer, Herbert, **4:**477, **7:**872
sperm, **5:**663
spider, **6:***713*
Spiegel, David, **6:**807
spinning wheel, **9:**1190
spiny spurge, **4:***475*
spiritism. *See* animism
Spirit Lake, **1:***74*
spirogyra, **3:***319*
Spitsbergen (Norway), **10:**1379, 1380

spontaneous generation, **4:**473
sport fishing. *See* fishing, sport
spotted owl, **4:**533, 546, **7:**869
Sri Lanka
  dengue hemorrhagic fever, **3:**286
  family planning program, **4:**483
Standard Oil Trust, **7:**919
Standard State Soil Conservation Districts
    Law, **9:**1150
staphylococcus, **1:**94–95, **3:**279
starfish, **6:**740
stargazing, **6:**773
starling, **4:**479
Starr, Chauncey, *as contributor,* **8:**1096–1097
stars, hydrogen and, **5:**678
START Regional Research Networks, **6:**720–721
State Department, U.S., **10:**1314
State Environmental Policy Acts, **4:**412
Statue of Liberty, **7:**955
steam engine, **9:**1191–1192
steel mills. *See* iron and steel
Steen, Harold K., *as contributor,* **4:**540–543
Stegner, Wallace, **6:**777
Steiner, Frederick, *as contributor,* **6:**748–753
Steno, Nicholas, **7:**942
steppe, **1:**119
sterilization, **2:**223, **4:**484, **8:**1014
Stewart, Omer C., **4:**493
Stirling cycle engine, **2:**202
Stockholm (Sweden), **10:***1293*
  metro system, **9:**1246
  suburban planned communities, **10:**1302
Stockholm Conference (1972), **3:**304, 305,
    307, **4:**414, 418, 426, 427, 431, **8:**1076,
    9:1212, **10:**1290, 1294, *1295*, 1328
Stoeckle, Andrew, *as contributor,* **5:**610–611
Stoel, Thomas B., Jr., *as contributor,* **4:**426–431
Stone, Christopher, **5:**665
Stonehenge, **9:**1253
Storm King Mountain (N.Y.), **1:***26*, 28,
    4:422, 423
storms, **6:**818–819
  hurricane, **7:**854, 856, 913
  tornado, **6:**818–819, **7:**855
  tsunami, **7:**855, 910
  See also flood
Stoss, Frederick W., *as contributor,*
    11:1431–1458
stoves, fuel-efficient, **5:**569
Strait of Belle Isle, **10:**1379
Strait of Gibraltar, **7:**910
Strasser, Stephen, **5:**650
Strassmann, Fritz, **7:**892
stratification, lake, **6:**746–747, **10:**1365–1366
stratigraphy, **7:**942–943
stratosphere, **1:**82, *83*, 86–87
streetcars, **9:**1246
street trees, **4:**537
streptococci, **1:**94, 95
Strier, Karen B., *as contributor,* **1:**48–53
strip mining, **2:***153*, 213
Strong, Maurice, **4:***426*
subdermal implant, **4:**484
submarines, **7:**913
substance abuse, **6:**803
Substance Abuse and Mental Health
    Services Administration, **8:**1030
suburb, **6:***763*, 765, **9:1160–1163**
  automobile use, **9:**1161–1162, 1242

Page numbers in **boldface type** indicate full articles on a subject. Page numbers in *italic type* indicate illustrations or other graphics. All numbers preceding colons indicate volume numbers.

tern, arctic, **7:**868
TERRA (remote sensing spacecraft), **3:**308–310, 311, **8:**1079
terrestrial biome, **1:**118, 119, 120–122
territoriality, **9:1206–1207**
Tester, Jefferson W., *as contributor*, **3:**376–381
tetraethyl lead, **6:**703, **8:**982
Texas
    Colorado River, **9:***1201*
    Galveston flood, **7:**856
    irrigation, **4:**502, *504*
    Lady Bird Johnson Wildflower Research Society, **10:**1385
    mineral production, **7:**841
    natural gas reserves, **7:**858
    organic farming, 1174
    planned communities, **10:**1302
    ranching, **8:**1048–1049
    rangeland, **8:**1053
    savanna, **1:**122
    sport hunters, **5:**675
    toxic wastes, **9:***1230*
Texas Railroad Commission, **7:**919
Thai Forest Monks, **8:**1068, 1069
Thailand
    agroforestry, **1:***14*
    Buddhist monks, **8:**1069, *1071*
    demographic transition, **8:**1009
    roads and bridges, **9:***1240*
    woodcarving factory, **6:***701*
Thales, **3:**354
Thames River
    cholera contamination, **4:**456–457
    ecosystem restoration, **3:**347, **8:**1086, *1088*
Theory of Moral Sentiments, The (Smith), **9:**1138–1139
Theory of the Earth (Lyell), **6:**784
thermal environment, **9:1208–1209**
thermal pollution, **9:1210–1211**
thermal vents. *See* hydrothermal vents
thermodynamics, laws of, **3:**316, 382, 392, **4:**517, 519
    and energy crisis, **3:**394
    statement of, **3:**393
thermokarst, **10:**1280
thermoplastics, **8:**1000–1001
thermosets, **8:**1001
thermosphere, **1:**82, *83*
Think Globally, Act Locally, **9:1212–1213**
Thomas, Lee, **4:**451
Thomas, Lewis, **6:**780
Thompson, Benjamin, **3:**392
Thomson, William (Lord Kelvin), **5:**584
Thoreau, Henry David, **2:**214, **4:**422, 464, 545, **6:**709, 737, **9:1214–1215**
    as abolitionist, **9:**1238
    as Emerson's disciple, **3:**362, **6:**775, **9:**1214
    literature by, **6:**775
    Muir compared with, **7:**843
    on nature as key to self-discovery, **7:**952, 953
    as transcendentalist, **9:**1236, 1238, 1239
Thoreau, John, **9:**1214
Three Gorges Project, **5:**682, 685, **9:**1200, 1201, **1216–1219**
Three Mile Island, **2:**205, **6:**705, **7:**895, 897, **9:1220–1221**
Throop (Pa.), **4:***453*

thunderstorms, **6:**818
Tiananmen Square (Beijing), **5:***664–665*
Tibet, **8:**1068
Tibetan Plateau, **9:**1200
tick, **3:**289, **6:**717, **7:**960
tidal energy. *See* energy, tidal
Tifernine Dunes (West Algeria), **2:***263*
Tigris River and Valley, **2:**144, **5:**600, **6:**751, **9:**1199
Tikal (Guatemala), **5:***589*
Titicaca, Lake, **8:***1077*
Titusville (Pa.), **8:**980
tobacco, **9:1222–1223**, 1226
    air pollution, **1:**18–19, **5:**631
    as carcinogen, **2:**147, *162*
    cigarette air pollution, **1:**18–19, **9:**1222–1223
    lung cancer risk, **1:**21–22, 77
    smoking death rates, **5:**631, 632
Tokyo (Japan)
    air pollution, **1:**18
    electronics store, **3:***387*
    suburban planned communities, **10:**1303
    urban transit system, **10:**1303
Toltec burial grounds, **1:***53*
Tomek, F. F., **10:**1390
Tongass National Forest, **4:***540*
tools. *See* technology
topography
    gross primary production, **7:***913*
    land and sea, **7:***908–909*
tornado, **6:***818–819*, **7:**855
Toronto (Canada), **8:**1004
totem pole, **1:***49*
tourism. *See* travel and tourism industry
Tower of Babel, **10:***1396*, 1397
town planning, **6:**761
toxic metal, **5:**625, **9:1224–1227**
    carbon monoxide, **2:**161
    mercury as, **6:**814–815
toxicology, **4:**452, **9:1228–1231**
    air pollutants, **2:**179
    Bhopal gas-leak disaster, **1:**108–109
    carbon monoxide, **2:**160–161
    carcinogens, **2:**162–163
    chemical industry, **2:**174–175
    epidemiology, **4:**457
    lead, **6:**766–767
    nitrate/nitrite, **7:**882
    occupational safety, **7:**874, 903
    pest control and pesticides, **2:**167, **6:**704, 710–711, 717, **7:**966, 967–968
    public health, **7:**874
    red tide, **1:**25
    water pollutants, **2:**180
    wild food plants, **8:**998
    See also DDT; food contamination
Toxics Release Inventory, **9:**1169
Toxic Substances Control Act, **2:**147, **4:**424, 450, **6:**744, 745
tradable permit, **9:1232–1233**
    pollution trading, **8:**1006–1007
Trager, William, *as contributor*, **7:**944–947, **9:**1182–1185
Train, Russell, **4:**449
Trans-Alaskan Pipeline, **9:1234–1235**
transcendentalism, **2:**214, **3:**362, 363, **9:1236–1239**

transformational evolution, **4:**473
transgenes, **5:**579
transpiration, **3:**301, **10:**1356, 1357
transportation, **9:**1190, **1240–1243**
    automobile, **1:**90–93
    canals as, **2:**145
    energy sources, **3:**380–381, 385–386, 387–388
    of farm produce, **6:**765, 823
    noise pollution, **7:**890, 891
    railroad coal cars, **2:***194*, 197
    road and highway, **8:**1100–1105
    urban, **10:**1302–1304
    zero-emission vehicle, **10:**1392–1395
    See also travel and tourism industry
transportation, commuter, **8:**1004, *1005*, **9:1244–1245**
    congestion, **9:**1241
    rail systems, **9:**1246–1247
    road and highway, **8:**1105
    suburbs and, **9:**1162, 1163, 1244, *1245*
    urban development and, **10:**1303–1304
transportation, rail, **8:**1101, **9:**1192, 1242–1243, **1246–1247**
    coal transport, **2:***194*, 197
    electric-powered trains, **3:**381, 385–386, 388
    high-speed bullet trains, **10:**130
    land use, **6:**752, 754, *764*, 765
    migrant labor, **6:**823, 825, 827
    suburban development and, **9:**1161, 1163, *1245*
    urbanization and, **10:**1307
Transportation Department, U.S., **4:**422, **5:**566, **8:**1101, **10:**1314
Transportation Equity Act for the 21st Century, **8:**1102
travel and tourism industry, **7:**915, **9:1248–1253**, **10:**1274
    Antarctica, **1:**46
    Aswan City, **1:**81
    ecotourism, **1:**101, 113, **9:**1249, 1252–1253
    Lake Baikal, **1:**101
    outdoor recreation, **8:**1057–1059
    rangeland settings, **8:**1052
tree farming, **9:***1171*, **1254–1255**
    agroforestry, **1:**14–15
    Judaic tradition, **8:**1073
    See also agroforestry
tree rings, **2:**187
Trenton (N.C.), **8:**1035
Trésaguet, Pierre-Marie-Jérôme, **8:**1101
Trinidad, **7:***970–971*
trophic level, **4:**516, 519
trophic niche, **7:**877
tropical dry forest, **9:1256–1259**
tropical rain forest, **3:***405*, **4:**438, **6:***720*, 737, 738, **9:**1197, **10:1264–1271**, 1274
    biome, **1:**120
    clear-cutting effects, **2:**182
    climate change effects, **2:**187
    debt-for-nature swap, **2:***247*
    deforestation, **2:**158, 252, *253*, 254–255
    ecological damage, **2:**128, 217
    fire deforestation, **4:**492
    international programs, **4:**431, 446, **10:**1296
    as natural habitat, **7:**864

Page numbers in **boldface type** indicate full articles on a subject. Page numbers in *italic type* indicate illustrations or other graphics. All numbers preceding colons indicate volume numbers.

nuclear power, **7:**892–893, 896–897
occupational safety and health measures, **2:**174, 175, **7:**874, 899, 901–902, 903
public health practice, **8:**1031–1032
soil conservation programs, **1:**106–107, **9:**1147, *1148*, 1157
Strategic Petroleum Reserve, **8:**981
technology-forcing regulation, **9:**1194–1195
Tennessee Valley Authority, **9:**1202–1203
waste management, **10:**1341–1342, 1343
water pollution control, **10:**1368–1369
water quality regulation, **2:**180–181
wildlife refuge system, **7:**850–853
See also environmental law *headings*; Environmental Protection Agency; *other specific departments and agencies*
Utah, **6:***751*
Bryce Canyon, **7:***953*
Great Salt Lake, **6:**746
mineral production, **7:**841
rangeland, **8:**1049, 1053
Utell, Mark J., *as contributor*, **2:**160–161
uterine cancer, **2:**162
utilitarianism, **2:**214, **6:**791
utopianism, **10:**1316–1317
Uzbekistan, **1:**60
solar energy, **3:***393*, **5:***554*

vaccines, **1:**114, **2:**132, 134, 149
demographic transition effects, **8:**1011
public health efforts, **8:**1028–1029
virus, **10:**1321
vacuum freezing, **2:**259
Vacquier, Victor, **7:**907
Vaiont Dam, **4:**510
Valdez (Alaska), **9:**1234, 1235
Valentine (La.), **4:***425*
Vancouver (Canada), **6:***749*, **9:***1167*
Vancouver Conference (1976), **10:**1290–1291
Vancouver Island (Canada), **3:**307
Vandalia (Ill.), **8:**1100
Van de Walle, Etienne, *as contributor*, **6:**788–789
Van Doren, Carlton S., *as contributor*, **8:**1056–1059
Van Schilfgaarde, Jan, *as contributor*, **6:**728–732
Van Vliet, William, *as contributor*, **5:**642–645
vapor. *See* water cycle
variational evolution, **4:**473
vasectomy, **2:**223
Vasuki, N. C., *as contributor*, **10:**1336–1340
Vaux, Calvert, **4:**464, **6:**758, **7:**916, 949
Vaux Le Vicomte (France), **6:**757
Vedic tradition, **8:**1068–1070
vegetarianism, **1:***40*
vegetation
as ant diet, **9:**1144
biome, **1:**118
dryland, **3:**271, 273
ecosystem development, **3:**344
historical changes in, **3:**321–323
ozone production, **9:**1142
rangeland, **8:**1051–1054
temperate rain forest, **9:***1196*, 1197
tropical dry forest, **9:***1256*, 1257–1258
tropical rain forest layers, **10:**1264–1265

tropical river basin, **10:**1272
tundra, **10:**1278, 1279, 1280
water hyacinth, **10:**1362–1363
wildflower, **8:***986*, **10:**1384–1385
See also forest *headings*; wildlife
Velikhov, Evguenji, **8:**1065
Venezuela
OPEC membership, **7:**919, 920, 923
petroleum, **8:**980
ranches, **8:**1048
savanna grassland, **1:**119
tropical dry forest, **9:**1256
Venice (Italy), flood, **4:**510
ventilation, **1:**21
Venus (planet), **7:***927*
Venus flytrap, **3:***372*, 373
Vermont, used oil collection tanks, **5:**624
Verona (Italy), **3:**306
Versailles (France), **6:**757
Victoria Island (Canada), **10:**279
Vienna (Austria), **9:**1164
Vienna Convention for the Protection of the Ozone Layer (1985), **4:**427–428
Vietnam
anti-dengue fever project, **3:***288*
percent of underweight children, **5:**673
refugees, **6:**826
Vietnam War, **3:**283, **10:**1325, 1326, 1328
Villa d'Este (Rome), **6:**758
Villa Lante (Bagnaia), **6:**758
vinyl chloride, **2:**179
Virginia
Chincoteague National Wildlife Refuge, **7:***867*
fuel tankers, **7:***858–859*
Monticello, **6:**758, 775
Norfolk waterfront renovation, **10:**1303
Reston planned community, **5:**609, **9:**1163, **10:**1302
Shenandoah National Park, **9:**1142–1143
TVA mandate in, **9:**1202
urban growth and development, **10:**1304–1305
Virji, Hassan, *as contributor*, **6:**718–721
virus, **5:**633, **10:**1318–1321
asexually reproduced clones, **2:**190
carcinogenic, **2:**146, 149
diseases, **3:**284–289
germ warfare, **5:**590–591
HIV and AIDS, **1:**16–17
parasitism, **7:**944
pest control, **7:**963
visual art. *See* art esthetics
vitamins, **3:**278, **6:**724
Vogt, William, **2:**216, **6:**777
volcanic eruption
ash, **1:**78, 79
climate effects, **2:**185
coral reef effects, **2:**227
geologic formation, **5:***582*, 585, 587
greenhouse gases, **1:***87*
as natural disaster, **7:**855, 857
nitrogen cycle and, **7:**886, *887*
oceanic, **7:**907, *910*
sulfur, **9:***1166*, 1167
Vuchic, Vukan R., *as contributor*, **9:**1246–1247

Walbridge, Mark R., *as contributor*, **6:**828–833

Walden (Thoreau), **2:**214, **3:**362, **6:**737, 775, **7:**953, **9:**1214–1215
Walden Pond, **6:**775, **9:**1214, *1215*, 1236
Wales, **4:***573*, **5:**583
Walker, H. Jesse, *as contributor*, **1:**70–73, **9:**1132–1135
Walker, James C. G., *as contributor*, **7:**938–941
Wallace, Alfred, **2:**239, **10:**1322–1323
Wallace, Henry, **2:**217
Waller, William T., **3:**275
Walsh, Barry Walden, *as contributor*, **8:**990–991
Walsh-Healey Public Contracts Acts, **7:**899, 901
Walth, Brent, *as contributor*, **6:**794–795
Walton, Izaak, **4:**496
Walvis Bay, **5:***620*
Wang, Herbert Han-Pu, *as contributor*, **5:**610–611
war and military activity, **10:**1324–1327
germ warfare, **5:**590–591
hunger created by, **5:**670–671
migrants from, **6:**825–826
nuclear weapons, **7:**892–893, 895, 913
oceans and, **7:**913
oil prices and, **7:**920–923
weapons development, **9:***1188*, 1190
warblers, **2:**137
Ward, Barbara, **6:**778, **10:**1290, **1328–1329**
Ware, George W., *as contributor*, **7:**960–965
Warren, Karen J., *as contributor*, **3:**312
Washington, George, **2:**213
Washington, D.C.
AIDS quilt, **1:***17*
asphalt paving, **8:**1100
automobile emissions, **9:**1142
Doxiades urban planning, **3:**296
Earth Day rally, **3:**306, *307*
historical sanitation problems, **10:**1332
metro system, **6:**765, **9:***1162*, 1242, 1246
Olmsted project, **7:**917
Smithsonian Institution, **6:**790, **10:**1314–1315
suburban commuters, **9:***1162*
Supreme Court building, **4:***432*
transportation corridors and redevelopment, **8:**1104, **9:**1243, **10:**1304
Washington State
Benson Creek landslide, **10:***1371*
Cascades National Park, **7:***952*
Eagle Harbor, **9:***1168*
Gifford Pinchot National Forest, **6:***739*, **8:***991*
Mount Olympus, **8:**1110
Olympic National Forest, **4:***545*
organic farm, **9:***1170*
Palouse Prairie, **8:***1053*
Pleistocene Era flood, **5:**596
rangeland, **8:**1049
Spirit Lake, **1:***74*
See also Seattle
wasp, **9:**1144, 1146–1147
waste, **10:**1330–1335
as archeological artifact source, **1:***64*
biomass, **8:**1083–1084
generation, recovery, and disposal, **10:***1340*
as water polluter, **10:***1358*, 1364

Page numbers in **boldface type** indicate full articles on a subject. Page numbers in *italic type* indicate illustrations or other graphics. All numbers preceding colons indicate volume numbers.